Problem-Solving Methods in Combinatorics

D1455194

Birkhäuser

Pablo Soberón

Problem-Solving Methods in Combinatorics

An Approach to Olympiad Problems

Birkhäuser

Pablo Soberón
Department of Mathematics
University College London
London, UK

ISBN 978-3-0348-0596-4   ISBN 978-3-0348-0597-1 (eBook)
DOI 10.1007/978-3-0348-0597-1
Springer Basel Heidelberg New York Dordrecht London

Library of Congress Control Number: 2013934547

Mathematics Subject Classification: 05-01, 97K20

Printed on acid-free paper

Springer Basel is part of Springer Science+Business Media (www.birkhauser-science.com)

Contents

Introduction

Every year there is at least one combinatorics problem in each of the major mathematical olympiads of international level. These problems have the common trait of needing a very high level of wit and creativity to find a solution. Even in the most recent competitions there are difficult problems that can be solved almost completely by a single brilliant idea. However, to be able to attack these problems with comfort it is necessary to have faced problems of similar difficulty previously and to have a good knowledge of the techniques that are commonly used to solve them.

I write this book with two purposes in mind. The first is to explain the tools and tricks necessary to solve almost any combinatorics problems in international olympiads, with clear examples of how they are used. The second way to offer to the olympic students (and other interested readers) an ample list of problems with hints and solutions. This book may be used for training purposes in mathematical olympiads or as part of a course in combinatorics.

Despite the fact that this book is self-contained, previous contact with combinatorics is advisable in order to grasp the concepts with ease. In the section "Further reading" we suggest material for this purpose. Reading this would be especially useful for familiarizing with the notation and basic ideas. It is also required to have a basic understanding of congruences in number theory.

The book is divided into 7 chapters of theory, a chapter of hints and a chapter of solutions. The theory chapters provide examples and exercises along the text and end with a problems section. In total there are 127 problems in the book. The purpose of the exercises is for the reader to start using the ideas of the chapter. There are no hints nor solutions for the exercises, as they are (almost always) easier than the problems. All examples and problems have solutions.

Although the text is intended to be read in the order it is presented, it is possible to read it following the diagram below.

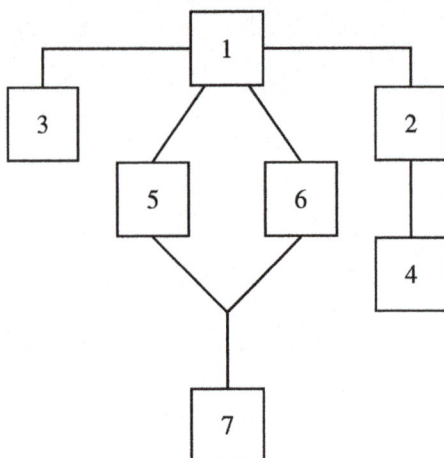

Many of the problems and examples of this book have appeared in mathematical contests, so I have tried to provide references to their first appearance. I apologize if any are missing or just plain wrong. At the end of the book, where the notation is explained, there is a list of abbreviations that were used to make these references.

I would like to thank Radmila Bulajich for both teaching me the necessary LaTeX to write this book and to motivating me to do so. Without her support this project would have never been brought to completion. I would also like to thank the extensive corrections and comments by Leonardo Martínez Sandoval as well as the observations by Adrián González-Casanova Soberón and Carla Márquez Luna. Their help shaped this book to its present form.

Pablo Soberón

For the Student

Even though a large number of problems in combinatorics have a quick and/or easy solution, that does not mean the problem one has to solve is not hard. Many times the difficulty of a problem in combinatorics lies in the fact that the idea that works is very well "hidden". Due to this the only way to really learn combinatorics is solving many problems, rather than reading a lot of theory. This practice is precisely what teaches you how to look for these creative or hidden ideas.

As you read the theory you will find exercises and examples. It is important to do them as they come along, since many of them are an important part of the theory and will be used repeatedly later on. When you find an example, try to solve it by yourself before reading the solution. This is the only way to see the complications that particular problem brings. By doing this you will have an easier time understanding why the ideas in the solution work and why they should be natural.

At the end of each chapter there is a section with problems. These problems are meant to be harder than the exercises and they have hints and solutions at the end of the book. Many problems are of an international competition level, so you should not become discouraged if you find a particularly difficult one.

Some problems have more than one solution. This does not mean that these solutions are the only ones. Chase your own ideas and you will probably find solutions that differ from the ones presented in this book (and perhaps are even better). Thus, when you reach a problems section, do not restrict yourself to the tools shown in that chapter.

Finally, my main objective when writing this book was not simply to teach combinatorics, but for the book to be enjoyed by its readers. Take it easy. Remember that solving problems is a discipline that is learned with constant practice and leads to a great sense of satisfaction.

First Concepts

1

1.1   Sets and First Countings

A **set** is defined to be a "collection of elements contained within a whole". That is, a set is defined only by its elements, which can also be sets. When we write $\{a, b, c\} = S$, we mean that S is the set that has the elements a, b, c. Two sets are equal if (and only if) they have the same elements. To indicate that a is an element of S, we write $a \in S$. In a set we never repeat elements, they can just be in the set or not be there.

Given a set A and a property ψ, we denote by $\{a \in A \mid \psi(a)\}$ the set of all elements of A that satisfy property ψ. For example, $\{a \in \mathbb{R} \mid a > 2\}$ is the set of all real numbers that are greater than 2. By convention, there is an empty set. That is, a set with no elements. One usually denotes that set with the symbol $\emptyset$.

When we say that a set has to be contained within a whole, we mean that it should not be "too big". This is a very technical detail, but there are collections (such as the collection of all sets) that bring difficulties if we consider them as sets. Throughout this book we will never face this kind of difficulties, even when we talk about infinite sets. However, it is important to know that there are collections that are not sets.[1]

Given two sets A and B, we say that A is a **subset** of B, or that A **is contained** in B, if every element of A is an element of B. We denote this by $A \subset B$. For example $\{1, 2\} \subset \{1, \{1, 3\}, 2\}$ since 1 and 2 are elements of the second set. However, $\{1, 3\} \not\subset \{1, \{1, 3\}, 2\}$ since 3 is not an element in the second set.

Given two sets A and B, we can use them to generate other sets.

- $A \cap B$, called the **intersection** of A and B. This is the set that consists of all the elements that are in both A and B.
- $A \cup B$, called the **union** of A and B. This is the set that consists all the elements that are in at least one of A or B.

---

[1]If you want an example of this, consider **Russell's paradox**. Let S be the collection of all sets that do not contain themselves as elements, i.e., all sets A such that $A \notin A$. If S is considered as a set, then $S \in S$ if and only if $S \notin S$, a contradiction!

P. Soberón, *Problem-Solving Methods in Combinatorics*,
DOI 10.1007/978-3-0348-0597-1_1, © Springer Basel 2013

- $\mathcal{P}(A)$, called the **power set** of A. This is the set made by all subsets of A. Using the notation above, we would write $\mathcal{P}(A) = \{B \mid B \subset A\}$.

In general we have that $A \subset A$ and $\emptyset \subset A$ for any set A, so they are elements of $\mathcal{P}(A)$. Notice that $\mathcal{P}(\emptyset)$ is not $\emptyset$, but rather $\{\emptyset\}$. We say that two sets A and B are disjoint if $A \cap B = \emptyset$. In other words, they are disjoint if they have no elements in common. Given a set A, $|A|$ denotes the number of elements of A. This number is called the **cardinality** of A.

Example 1.1.1 Prove that $A \cap B$ is the largest set that is contained in A and in B.

Solution Note that by the definition of $A \cap B$, it is contained in both A and B. Consider any set C that is contained both in A and B. Given any $x \in C$, we know that $x \in A$ since C is contained in A and $x \in B$ since C is contained in B. Thus, $x \in A \cap B$. Since this works for every $x \in C$, we have that $C \subset A \cap B$, as we wanted.   □

Exercise 1.1.2 Prove that $A \cup B$ is the smallest set that contains both A and B.

Exercise 1.1.3 Prove that if $A \subset B$, then $\mathcal{P}(A) \subset \mathcal{P}(B)$.

Exercise 1.1.4 Prove that if $\mathcal{P}(A) = \mathcal{P}(B)$, then $A = B$.

The concepts of union and intersection can be extended a bit more. If C is a set of sets, then $\bigcup C$ is the set made by all the elements that are in at least one set of C, and $\bigcap C$ is the set that consists of all the elements that are in every set of C. For example, $\bigcap \{A, B\} = A \cap B$.

These are the concepts that we are going to work with. We are going to be interested in how they behave. Combinatorics is the branch of mathematics that studies finite or discrete sets and structures (such a graphs). For example, counting or enumerating sets or events are part of combinatorics. Every time that a question of the type "In how many ways can this happen?" arises, the arguments that are going to be used for the solution are part of combinatorics. Sets are the basic objects we are going to be counting. This will be done either by their properties or by the events that they represent. To do these countings, there are two basic laws we have to follow.

Law of the sum   If an event A can happen in m different ways and an event B can happen in n different ways, then the number of ways in which A or B can happen is $m + n$.

Law of the product   If an event A can happen in m different ways and an event B can happen in n different ways, then the number of ways in which A and then B can happen is mn.

Example 1.1.5 Given non-negative integers $k \leq n$, how many ordered lists of k different elements can be made if there are n different elements available?

Solution Notice that to choose the first element, there are n options. Then, to choose the second element, there are $n - 1$ options. By the law of the product, to choose the first two elements there are $n(n - 1)$ options. To choose the third element there are $n - 2$ options, so to choose the first three elements there are $n(n - 1)(n - 2)$ possible ways. If we continue this way, to choose the first k elements (in order) there are $n(n - 1)(n - 2) \cdots (n - k + 1)$ different ways. □

If in the previous example $n = k$, then we see that the number of ways to order the n elements is $n \cdot (n - 1) \cdot (n - 2) \cdots 2 \cdot 1$. This is the number of ways to order all the elements of the original set. This number is denoted by $n!$, which is read as "n **factorial**". We define $0!$ as 1. An ordering of a list is also called a **permutation** of the list. In the following chapters we will study permutations from another point of view.

Example 1.1.6 If A is a set with n elements, how many subsets does it have?

Solution To solve this example the event we want to count is that of forming a subset of A. We can form it by choosing one element at a time. That is, for each element of A we choose whether it is going to be in the subset or not. For every element there are 2 options (to be included in the subset or not), so using the law of the product $n - 1$ times we obtain that the number of subsets of A is 2^n. □

In other words, if $|A| = n$, then $|\mathcal{P}(A)| = 2^n$.

Proposition 1.1.7 *The number of subsets of k elements of a set with n elements is* $\frac{n!}{k!(n-k)!}$.

Proof To see this, notice that when we count the number of lists of k elements, every subset of size k is counted once for every way of ordering its elements. The number of lists of k elements is $n(n - 1) \cdots (n - k + 1)$ and the number of ways to order k elements in a list is $k!$. Thus the number we are looking for is $\frac{n(n-1)(n-2)\cdots(n-k+1)}{k!} = \frac{n!}{k!(n-k)!}$. □

These numbers are also denoted by $\binom{n}{k}$, C_k^n or C_n^k and are known as the **binomial coefficients**. $\binom{n}{k}$ is usually read as "n choose k". They are given this name because, when we expand $(a + b)^n$, these numbers appear in the coefficients. In a more explicit way:

Theorem 1.1.8 (Newton) *Let n be a positive integer. Then*

$$(a + b)^n = \binom{n}{0}a^n + \binom{n}{1}a^{n-1}b + \binom{n}{2}a^{n-2}b^2 + \cdots + \binom{n}{n}b^n = \sum_{k=0}^{n} \binom{n}{k}a^{n-k}b^k.$$

Proof First we have to notice that

$$(a+b)^n = \underbrace{(a+b)(a+b)\cdots(a+b)}_{n \text{ times}}.$$

That is, in each factor $(a+b)$ of the product we have to choose either a or b. Since there are n factors, every term in the result is of the form $a^r b^s$ with $r + s = n$. The term $a^{n-k} b^k$ appears once for every way to choose k times the term b in the product. There are $\binom{n}{k}$ ways to do this, which is the number we were looking for.

We are going to set $\binom{n}{k} = 0$ if k is greater than n or is negative. $\square$

Example 1.1.9 In a set there are n red objects and m blue objects. How many pairs of elements of the same color can be made?

Solution There are two cases: that the pair of objects is red or the pair of objects is blue. In the first case there are $\binom{n}{2}$ ways to choose the pair and in the second there are $\binom{m}{2}$ ways to choose the pair. By the law of the sum there are in total $\binom{n}{2} + \binom{m}{2}$ possible pairs. $\square$

Exercise 1.1.10 Prove that $\binom{n}{k} = \binom{n}{n-k}$.

Exercise 1.1.11 (Pascal's formula)   Prove that $\binom{n}{k} + \binom{n}{k+1} = \binom{n+1}{k+1}$.

Exercise 1.1.12 Prove that $2^n = \binom{n}{0} + \binom{n}{1} + \cdots + \binom{n}{n}$.

Exercise 1.1.13 Prove that if $n \geq 1$, then

$$0 = \binom{n}{0} - \binom{n}{1} + \binom{n}{2} - \binom{n}{3} + \cdots + (-1)^n \binom{n}{n}.$$

Exercise 1.1.14 Prove that $\binom{n+1}{k+1} = \frac{n+1}{k+1} \binom{n}{k}$.

Exercise 1.1.15 Prove that $(n - 2k)\binom{n}{k} = n\left[\binom{n-1}{k} - \binom{n-1}{k-1}\right]$.

Exercise 1.1.16 Using Exercise 1.1.15 prove that[2]

$$\binom{n}{0} < \binom{n}{1} < \binom{n}{2} < \cdots < \binom{n}{\lfloor \frac{n}{2} \rfloor}.$$

Exercise 1.1.17 Prove that $\binom{n+1}{3} - \binom{n-1}{3} = (n-1)^2$.

----

[2] $\lfloor x \rfloor$ denotes the greatest integer that is smaller than or equal to x.

Exercise 1.1.18 Prove that if $n > r > 0$, then $\binom{n}{r}^2 > \binom{n}{r+1}\binom{n}{r-1}$.

Exercise 1.1.19 Prove that if $a < b$, then $\binom{a}{2} + \binom{b}{2} \geq \binom{a+1}{2} + \binom{b-1}{2}$.

Exercise 1.1.20 Prove that if p is prime and $0 < k < p$, then $\binom{p}{k}$ is divisible by p.

1.2   Induction

Mathematical induction is a technique used to prove statements. The idea is similar
to making several domino pieces fall. If every piece is close enough to the previous
one and we make the first one fall, then they are all going to fall.

When we want to prove a statement about natural numbers, the idea is the same.
Here statements are going to involve a variable n which is a positive integer. We say
that $P(n)$ is the statement for a certain n.

To prove that $P(n)$ is true for all n, it is enough to prove the following:

- **Basis of induction:**
  Prove that $P(1)$ is true.
  (Verifying that the first piece of domino falls.)
- **Induction hypothesis:**
  Suppose that $P(n)$ is true (here we are thinking that n is a fixed integer).
- **Inductive step:**
  Prove that $P(n + 1)$ is true.
  (Showing the if the n-th piece of domino falls, then so does the $(n + 1)$-th.)

If these steps hold, then P is true for every positive integer n.

If we want to prove the statement starting on some n_0 that is not necessarily 1,
we only have to change the induction basis to: Prove that $P(n_0)$ is true.

Whenever we use induction, we want to reduce the proof of $P(n + 1)$ to $P(n)$.
Since we supposed that $P(n)$ was true (by the induction hypothesis), we are able to
finish. This method usually makes proofs much easier. Let us see some examples:

Example 1.2.1 (Gauss formula) Prove that if n is a positive integer, then $1 + 2 + 3 + \cdots + n = \frac{n(n+1)}{2}$.

Solution Take for $P(n)$ the assertion "$1 + 2 + 3 + \cdots + n = \frac{n(n+1)}{2}$".

- Basis of induction:
  If $n = 1$, then $1 = \frac{1(1+1)}{2}$, thus $P(1)$ is true.
- Induction hypothesis:
  Suppose that $P(n)$ is true.
- Inductive step:
  We want to prove that $P(n + 1)$ is true. To show that, notice that

$$1 + 2 + 3 + \cdots + (n + 1) = (1 + 2 + \cdots + n) + (n + 1).$$

By the induction hypothesis, this is equal to

$$\frac{n(n+1)}{2} + (n+1) = \frac{n(n+1) + 2(n+1)}{2} = \frac{(n+2)(n+1)}{2},$$

which is what we wanted to prove.   □

Example 1.2.2 Let k and n be non-negative integers, with $n \geq k$. Prove that $\binom{n+1}{k+1} = \binom{k}{k} + \binom{k+1}{k} + \cdots + \binom{n}{k}$.

Solution Consider $P(n)$ as "$\binom{n+1}{k+1} = \binom{k}{k} + \binom{k+1}{k} + \cdots + \binom{n}{k}$".

- Basis of induction:
  Since we want to prove the statement for $n \geq k$, the basis of induction should be when $n = k$. That is, to prove that

$$\binom{k}{k} = \binom{k+1}{k+1}.$$

  Since both parts are equal to 1, this is true.
- Induction Hypothesis:
  Suppose that $P(n)$ is true.
- Inductive step:
  We want to prove that $P(n+1)$ is true. That is, that

$$\binom{n+2}{k+1} = \binom{k}{k} + \binom{k+1}{k} + \cdots + \binom{n+1}{k}.$$

  Notice that

$$\binom{k}{k} + \binom{k+1}{k} + \cdots + \binom{n+1}{k} = \left[\binom{k}{k} + \binom{k+1}{k} + \cdots + \binom{n}{k}\right] + \binom{n+1}{k}.$$

  The first part is equal to $\binom{n+1}{k+1}$ by the induction hypothesis. So we want to prove that

$$\binom{n+2}{k+1} = \binom{n+1}{k+1} + \binom{n+1}{k}.$$

This is true by Pascal's formula (Exercise 1.1.11).   □

Here the induction was made on n. The statement really depends on k and n, so one might try using induction on k as well. However, in this case it would be much harder to reduce the problem to the previous case, so it is not a very good idea. Whenever the statement depends on more than one variable it is important to identify on which one it is easier to use induction.

Exercise 1.2.3 Prove that $1 + 3 + 5 + \cdots + (2n - 1) = n^2$ for every positive integer n.

Exercise 1.2.4 Prove that $1^2 + 2^2 + \cdots + n^2 = \frac{n(n+1)(2n+1)}{6}$ for every positive integer n.

Exercise 1.2.5 Prove that if q is a real number, $q \neq 1$ and n is a positive integer, then $1 + q + q^2 + \cdots + q^n = \frac{q^{n+1}-1}{q-1}$.

Exercise 1.2.6 Prove that if q is a real number, $q \neq 1$ and n is a positive integer, then $(1 + q)(1 + q^2)(1 + q^4) \cdots (1 + q^{2^n}) = \frac{1-q^{2^{n+1}}}{1-q}$.

Exercise 1.2.7 Prove that if q is a real number, $q \neq 1$ and n is a positive integer, then $1 + 2q + 3q^2 + \cdots + nq^{n-1} = \frac{1-(n+1)q^n+nq^{n+1}}{(1-q)^2}$.

Exercise 1.2.8 Prove that $\frac{1}{\sqrt{1}} + \frac{1}{\sqrt{2}} + \cdots + \frac{1}{\sqrt{n}} \geq \sqrt{n}$ for every positive integer n.

Exercise 1.2.9 Prove Theorem 1.1.8 by induction.

Exercise 1.2.10 Prove that if m, n are positive integers such that $0 \leq m \leq n$, then $\binom{n}{0} - \binom{n}{1} + \binom{n}{2} - \cdots + (-1)^m \binom{n}{m} = (-1)^m \binom{n-1}{m}$.
Note that this exercise implies Exercise 1.1.13.

The following exercises deal with Fibonacci's numbers; these are defined by Eqs. (6.3) and (6.4).

Exercise 1.2.11 Prove that $F_1 + F_2 + \cdots + F_n = F_{n+2} - 1$ for every positive integer n.

Exercise 1.2.12 Prove that $F_1 + F_3 + \cdots + F_{2n-1} = F_{2n}$ for every positive integer n.

Exercise 1.2.13 Prove that $F_{2n} = F_{n+1}^2 - F_{n-1}^2$ for every integer $n \geq 2$.

Exercise 1.2.14 Prove that $F_{2n+1} = F_{n+1}^2 + F_n^2$ for every integer $n \geq 2$.

Exercise 1.2.15 Prove that $F_1^2 + F_2^2 + F_3^2 + \cdots + F_n^2 = F_n F_{n+1}$ for every positive integer n.

In every proof by induction, we have to suppose that the previous case is valid. However, it may happen that the validity of the previous case is not enough. For such situations we can use **strong induction**. In this type of induction, instead of using that the previous case is valid, we are going to use that all previous cases are valid. That is, the induction hypothesis is that *for every $k \leq n$, $P(k)$ is true*. It can be proven that induction and strong induction are equivalent, but this will not be done in this book.

So far we have only seen how to use induction to prove formulas; now let us examine an example that is a bit more difficult (the first problem of international olympiad level we face!).

Example 1.2.16 (IMO 2002) Let n be a positive integer and S the set of points (x, y) in the plane, where x and y are non-negative integers such that $x + y < n$. The points of S are colored in red and blue so that if (x, y) is red, then (x', y') is red as long as $x' \leq x$ and $y' \leq y$. Let A be the number of ways to choose n blue points such that all their x-coordinates are different and let B be the number of ways to choose n blue points such that all their y-coordinates are different. Prove that $A = B$.

Proof Let a_k be the number of blue points with x-coordinate equal to k and b_k the number of blue points with y-coordinate equal to k. Using $n - 1$ times the law of the product, we have that $A = a_0 a_1 a_2 \cdots a_{n-1}$. In the same way we have that $B = b_0 b_1 b_2 \cdots b_{n-1}$. We are going to prove that the numbers $a_0, a_1, \ldots, a_{n-1}$ are a permutation of the numbers $b_0, b_1, \ldots, b_{n-1}$ using strong induction. By proving this, we will establish that their product is the same.

If $n = 1$, S consists of only one point. Then, a_0 and b_0 are both 1 or 0 depending on whether the point is painted blue or red, so $a_0 = b_0$. Suppose the assertion holds for every $k \leq n$ and we want to prove it for $n + 1$.

There are two cases: 1. every point (x, y) with $x + y = n$ is blue, or 2. at least one of them is red.

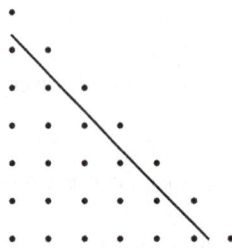
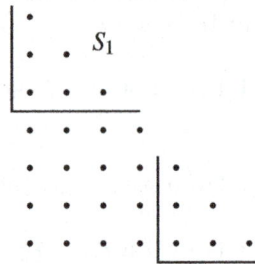

Case 1   Case 2

In case 1, let S' be the set of points (x, y) such that $x + y < n$ and let a'_k be the number of blue points in S' with k in the x-coordinate and b'_k the number of blue points in S' with k in the y-coordinate. Then $a'_0, a'_1, \ldots, a'_{n-1}$ is a permutation of $b'_0, b'_1, \ldots, b'_{n-1}$ (by the induction hypothesis). We also know that $a_n = b_n = 1$ and $a_k = a'_k + 1$, $b_k = b'_k + 1$ for every $k < n$. So $b_0, b_1, \ldots, b_n$ is a permutation of $a_0, a_1, \ldots, a_n$.

In case 2, let us suppose $(k, n - k)$ is red. Let S_1 be the set of points (x, y) with $x + y < n + 1$, $x < k$ and $y > n - k$. Notice that $a_0, a_1, \ldots, a_{k-1}$ and $b_{n-k+1}, b_{n-k+2}, \ldots, b_n$ are the a_i and b_i that would be associated with S_1. Thus, by the induction hypothesis, one is a permutation of the other. In the same way, $a_{k+1}, a_{k+2}, \ldots, a_n$ is also a permutation of $b_0, b_1, \ldots, b_{n-k-1}$. Since $a_k = b_{n-k} = 0$, we have that $a_0, a_1, \ldots, a_n$ is a permutation of $b_0, b_1, \ldots, b_n$.   □

The next example is significantly harder than the previous one, even though the solution seems simpler. This problem was the hardest one in the 2009 IMO and, if we consider the average number of points the students obtained, it is the second hardest problem to appear in an IMO (up to 2011). It is surprising that a problem with this level of difficulty can be solved using only the theory we have seen so far.

Example 1.2.17 (IMO 2009) Let $a_1, a_2, \ldots, a_n$ be different positive integers and M a set of $n - 1$ positive integers not containing the number $s = a_1 + a_2 + \cdots + a_n$. A grasshopper is going to jump along the real axis. It starts at the point 0 and makes n jumps to the right of lengths $a_1, a_2, \ldots, a_n$ in some order. Prove that the grasshopper can organize its jumps in such a way that it never falls in any point of M.

Solution We proceed by strong induction on n. For this we order the steps as $a_1 < a_2 < \cdots < a_n$ and the elements of M as $b_1 < b_2 < \cdots < b_{n-1}$. Let $s' = a_1 + a_2 + \cdots + a_{n-1}$. If we remove a_n and b_{n-1}, there are two cases.

- s' is not among the first $n - 2$ elements of M. In this case, by induction, we can order the first $n - 1$ jumps until we reach s'. If at any moment we fell on b_{n-1}, we change that last step for a_n and then we continue in any way to reach s. By induction we know that we have never fallen on $b_1, b_2, \ldots, b_{n-2}$. Also, if we had to use the change, since there are no elements of M after b_{n-1}, we do not have to worry about falling on a b_k in the rest of the jumps.
- s' is one of the first $n - 2$ elements of M. If this happens, then since $s' = s - a_n$ is in M, among the $2(n - 1)$ numbers of the form $s - a_i$, $s - a_i - a_n$ with $1 \leq i \leq n - 1$ there are at most $n - 2$ elements of M. If we look at the pairs of numbers $(s - a_i, s - a_i - a_n)$, since we have $n - 1$ of these pair and they contain at most $n - 2$ elements of M, there is a number a_i such that neither $s - a_i$, nor $s - a_i - a_n$ are in M. Notice that after $s - a_i - a_n$ we have s' and b_{n-1}, which are two elements of M. Therefore, there are at most $n - 2$ elements of M before $s - a_i - a_n$. Then, by the induction hypothesis, we can use the other $n - 2$ jumps to reach $s - a_i - a_n$, then use a_n and then use a_i to get to s without falling on a point of M.   □

1.3   Paths in Boards

Suppose we have a board of size $m \times n$ (m rows and n columns) divided into unit squares. How many paths are there on the sides of the squares that move only up or to the right and go from the bottom left corner to the top right one? (See Fig. 1.1.)

Solution Every path must go up m times and to the right n times. In total there must be $m + n$ steps. Consider an alphabet made only of the letters U and R. It is clear that for every path there is one and only one word of $m + n$ letters of this alphabet,

Fig. 1.1 Example of path if
$n = m = 5$

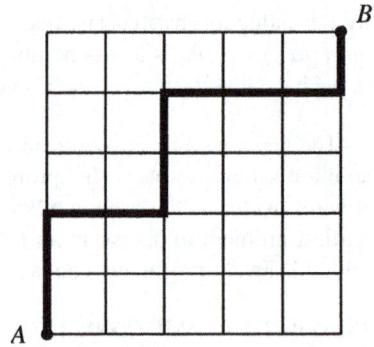

Fig. 1.2 Example with the
same path

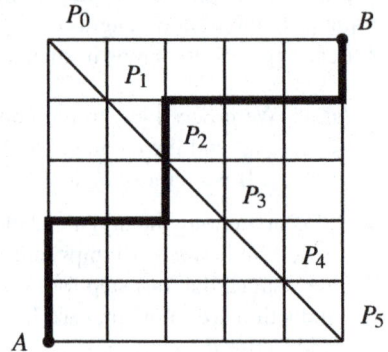

where U means to go up and R means to go to the right. For example, in the path of
the figure the word would be $(U, U, R, R, U, U, R, R, R, U)$.

To form a word with these properties, we only have to choose which are the m
places of the $m + n$ possible ones in which we are going to write the letter U. This
can be done in $\binom{m+n}{m}$ ways.   □

Proposition 1.3.1 *If n is a positive integer, then*

$$\binom{2n}{n} = \binom{n}{0}^2 + \binom{n}{1}^2 + \binom{n}{2}^2 + \cdots + \binom{n}{n}^2.$$

First Proof Consider an $n \times n$ board divided into unit squares. Then $\binom{2n}{n}$ is the
number of paths that move on the sides of the squares, move always up or to the
right and go from the bottom left corner to the top right one. Let $P_0, P_1, \ldots, P_n$ be
the points in the diagonal of the board that goes through the top left corner. (See
Fig. 1.2.)

If we want to go from the bottom left corner to the point P_k we have to move k
times to the right and $n - k$ times up. There are $\binom{n}{n-k}$ ways to reach this point. To
go from P_k to the top right corner of the board, we have to move $n - k$ times to the

right and k times up. There are $\binom{n}{k}$ ways of doing this. Thus the number of paths that go through P_k is $\binom{n}{n-k}\binom{n}{k} = \binom{n}{k}^2$ (by Exercise 1.1.10). Since every path must go through exactly one point of that diagonal, using the law of the sum n times we obtain the desired result. $\qquad\square$

Second Proof We prove the proposition by counting certain sets. Suppose we have a set of n blue balls and n red balls (all balls can be distinguished from each other). We want to count the number of subsets of n balls. We know there are $\binom{2n}{n}$ such subsets. Let us count how many sets there are such that k of the balls are red. For this we have to choose k red balls and $n-k$ blue ones, which gives $\binom{n}{k}\binom{n}{n-k} = \binom{n}{k}^2$ such sets. Since the number of red balls can vary from 0 to n, we get the desired result. $\qquad\square$

Exercise 1.3.2 Let p be a prime integer. Prove that $\binom{2p}{p} - 2$ is divisible by p^2 (recall Exercise 1.1.20).

Exercise 1.3.3 (Vandermonde's formula) Let n, m be non-negative integers such that $n \geq m$. Prove that

$$\binom{n+m}{m} = \binom{n}{0}\binom{m}{m} + \binom{n}{1}\binom{m}{m-1} + \binom{n}{2}\binom{m}{m-2} + \cdots + \binom{n}{m}\binom{m}{0}.$$

In some combinatorics problems it is very helpful to think in terms of paths in boards.

Let us see an example where we can use the proposition we just proved.

Example 1.3.4 (OIM 2005) Let n be a positive integer. On a line there are $2n$ marked points $A_1, A_2, \ldots, A_{2n}$. We are going to color the points blue or red in the following way: n disjoint circles are drawn with diameters of extremities A_i and A_j for some i, j. Every point A_k with $1 \leq k \leq n$ belongs to exactly one of these circles. The points are colored in such a way that the two points of a circle have the same color. Find the number of different colorings of the $2n$ points that can be obtained by changing the circles and the distribution of colors.

Solution Let $a_1, a_2, \ldots, a_{2n}$ be the points on the line, in that order. Notice that if two points are in the same circle, between them there must be an even number of points, and thus the circles join points with even index with points with odd index. Thus there is the same number of odd red points and even red points.

We are going to prove by induction on n that any coloring that has the same number of even red points and odd red points can be obtained by drawing circles. If $n = 1$, both points must have the same color, thus by drawing the circle that has their segment as diameter we are done. If there are two consecutive points with the same color, we can draw a circle that has them as diameter, and we are left with $2(n-1)$ points as indicated. These $2(n-1)$ points can be colored using circles (by the induction hypothesis). If there are no two consecutive points of the same color,

$$\bullet \ \bullet \ | \ \bullet \ \bullet \ || \ \bullet \ \bullet \ \bullet \ | \ \bullet \ | \ \bullet \ |||$$

Fig. 1.3 List of balls and delimiters that generates $(2, 2, 0, 3, 1, 1, 0, 0, 0)$

then all even points have one color and all odd points have another, which contradicts our hypothesis. Notice that the arrangement of the circles that gives a coloring does not need to be unique. For example, if all points are red, any arrangement of circles can give that coloring.

If we want to have k odd red points, we have $\binom{n}{k}$ ways to choose them, and $\binom{n}{k}$ ways to choose the even red points. Thus the total number of colorings is $\binom{n}{0}^2 + \binom{n}{1}^2 + \binom{n}{2}^2 + \cdots + \binom{n}{n}^2 = \binom{2n}{n}$. $\qquad\square$

1.4   A Couple of Tricks

In this section we mention two classic tricks that are used frequently in counting problems. The first one is to introduce **delimiters** to solve a problem. The second one is known as the inclusion-exclusion principle, and allows us to know the size of the union of some sets if we know the size of their intersections.

Example 1.4.1 Suppose that we have 9 identical balls and 9 distinguishable bins. We want to count the number of ways to distribute the balls in the bins (there may be several or no balls in the same bin). (See Fig. 1.3.)

Solution To count the different distributions, we are only interested in how many balls there are in the first bin, how many balls there are in the second bin, etc. In other words, the number of arrangements is the same as the number of ordered lists of integers $(n_1, n_2, \ldots, n_9)$ such that $n_i \geq 0$ for all i and $n_1 + n_2 + \cdots + n_9 = 9$. This is called a Diophantine equation (with coefficients 1).

Now consider 17 objects in a line. We are going to choose 8 of them to be the "delimiters". The other objects are going to represent the balls. We say that n_1 is the number of balls before the first delimiter, n_2 is the number of balls between the first delimiter and the second one, $\ldots$, n_9 is the number of balls after the eighth delimiter. It is clear that we obtain a list of the kind we were looking for, and also that any of such lists corresponds to an arrangement of 9 balls and 8 delimiters in a line. Thus the number we are looking for is $\binom{17}{8}$. $\qquad\square$

In the general problem when we want to distribute n identical object in m places, we can do it in $\binom{n+m-1}{m-1}$ ways, and that is equal to the number of lists $(n_1, n_2, \ldots, n_m)$ of non-negative integers such that $n_1 + n_2 + \cdots + n_m = n$.

Exercise 1.4.2 There are 10 boxes and 30 balls, of which 10 are green, 10 are blue and 10 are red. The balls of the same color are identical. In how many different ways can we distribute the 30 balls in the 10 boxes?

Exercise 1.4.3 Show that the number of lists $(b_1, b_2, \ldots, b_{m+1})$ of positive integers such that $b_1 + b_2 + \cdots + b_{m+1} = n + 1$ is $\binom{n}{m}$.

Example 1.4.4 Let m, k be positive integers. How many lists $(n_1, n_2, \ldots, n_k)$ of k numbers are there such that $0 \leq n_1 \leq n_2 \leq \cdots \leq n_k \leq m$?

First Solution (with delimiters) Consider an arrangement of m balls and k delimiters. If we label the delimiters from left to right, we can define n_i as the number of balls to the left of the i-th delimiter. It is now clear that the number of lists we are looking for is the same as the number of arrangements of m balls and k delimiters, which is $\binom{m+k}{k}$. □

Second Solution (without delimiters) Consider the list $(m_1, m_2, \ldots, m_k)$ defined by $m_i = n_i + i$. Then we have that $1 \leq m_1 < m_2 < \cdots < m_k \leq m + k$. For every list $(m_1, m_2, \ldots, m_k)$ that satisfies that condition we can construct $(n_1, n_2, \ldots, n_k)$. However, the m_i are k different numbers, and for every choice of k different numbers, they generate only one of such lists. Thus we only have to choose k numbers between 1 and $m + k$, for which there are only $\binom{m+k}{k}$ possible ways. □

Third Solution (using the previous example) Consider $b_1 = n_1$, $b_2 = n_2 - n_1$, $\ldots$, $b_k = n_k - n_{k-1}$, $b_{k+1} = m - n_k$. We know that $b_i \geq 0$ for all i and that $b_1 + b_2 + \cdots + b_{k+1} = m$. The number of lists $(b_1, b_2, \ldots, b_{k+1})$ with those properties is the same as the number of lists $(n_1, n_2, \ldots, n_k)$ we were looking for. By the previous example, we know there are $\binom{m+(k+1)-1}{(k+1)-1} = \binom{m+k}{k}$ such lists. In this solution, if we only define $b_1, b_2, \ldots, b_k$ and fix $n_k = j$, we know there are in total $\binom{j+k-1}{k-1}$ possible lists. Since j can go from 0 to m, we have that in total there are $\binom{k-1}{k-1} + \binom{k}{k-1} + \cdots + \binom{m+k-1}{k-1}$ lists. Using this and one of the previous solutions, we have a non-inductive solution to Example 1.2.2. □

Now let us introduce the **inclusion-exclusion** principle. For this, suppose we have three sets A, B, C and we want to find $|A \cup B \cup C|$. If we count the elements of each set, there are $|A| + |B| + |C|$. However, we have counted twice every element that is in two of them, so we must subtract $|A \cap B| + |B \cap C| + |C \cap A|$. However, by doing this we are no longer counting the elements that are in all three of them, so we must add $|A \cap B \cap C|$. In the end we have

$$|A \cup B \cup C| = |A| + |B| + |C| - \big(|A \cap B| + |B \cap C| + |C \cap A|\big) + |A \cap B \cap C|.$$

This line of reasoning can be extended to any finite number of sets.

Proposition 1.4.5 (Inclusion-exclusion principle) *Let $A_1, A_2, \ldots, A_n$ be sets. Then*

$$\left| \bigcup_{i=1}^{n} A_i \right| = \sum_{1 \leq i \leq n} |A_i| - \sum_{1 \leq i < j \leq n} |A_i \cap A_j| + \sum_{1 \leq i < j < k \leq n} |A_i \cap A_j \cap A_k| - \cdots$$
$$+ (-1)^{n+1} |A_1 \cap A_2 \cap A_3 \cap \cdots \cap A_n|.$$

That is, first we add the sizes of the sets, then we subtract the sizes of the inter-sections of the pairs of sets, then we add the sizes of the intersections of any three sets, etc.

Proof Let a be an element of $\bigcup_{i=1}^{n} A_i$ that is in exactly r sets. Then, in the right-hand side of the equation, we are first counting it $\binom{r}{1}$ times, then $\binom{r}{2}$ times, then $\binom{r}{3}$ times and so on. In total, we are counting it $\binom{r}{1} - \binom{r}{2} + \cdots + (-1)^{r+1}\binom{r}{r}$ times. By Exercise 1.1.13, this is equal to $\binom{r}{0} = 1$. Thus we are counting every element of $\bigcup_{i=1}^{n} A_i$ exactly once, as we wanted. □

The main problem with this formula is that, given its length and the number of sign changes, it is difficult to use. However, it still helps to solve several counting problems.

Example 1.4.6 Let n be a positive integer. Find the number of permutations of $(1, 2, \ldots, n)$ such that no number remains in its original place.

Solution To do this, first we are going to count the number of permutations where at least one number remains in its place. If A_i is the number of permutations where i remains in its place, we are looking for $|\bigcup_{i=1}^{n} A_i|$. Thus, according to the inclusion-exclusion principle, we must first add the permutations where a given number is fixed, then subtract the permutations where 2 given numbers are fixed and so on. To find a permutation that fixes k given elements, we only have to arrange the rest, which can be done in $(n - k)!$ ways. However, if we do this for every choice of k elements, we are counting $\binom{n}{k}(n - k)! = \frac{n!}{k!}$ permutations. Thus, the number of permutations that leave at least one point fixed is $\frac{n!}{1!} - \frac{n!}{2!} + \frac{n!}{3!} - \cdots + (-1)^{n+1}\frac{n!}{n!}$. Since there are $n! = \frac{n!}{0!}$ permutations in total, the number of permutations where no number is fixed is $n! - (\frac{n!}{1!} - \frac{n!}{2!} + \frac{n!}{3!} - \cdots + (-1)^{n+1}\frac{n!}{n!})$ which is equal to

$$n!\left[\frac{1}{0!} - \frac{1}{1!} + \frac{1}{2!} - \frac{1}{3!} + \cdots + (-1)^n\left(\frac{1}{n!}\right)\right].$$ □

Notice that in total there are $n!$ possible permutations. Thus, the portion of permutations that do not leave any number in its place is $\frac{1}{0!} - \frac{1}{1!} + \frac{1}{2!} - \frac{1}{3!} + \cdots + (-1)^n(\frac{1}{n!})$. An n grows, this sum gets closer to a certain number,[3] which is approximately 0.367879. That is, in general more than one third of the permutations leave no number in its place.

---

[3] The exact value of this number is $\frac{1}{e}$, where e is the famous **Euler constant**.

1.5   Problems

Problem 1.1 Find the number of ways to place 3 rooks on a 5×5 chess board so that no two of them attack each other.[4]

Problem 1.2 (Iceland 2009) A number of persons seat at a round table. It is known that there are 7 women who have a woman to their right and 12 women who have a man to their right. We know that 3 out of each 4 men have a woman to their right. How many people are seated at the table?

Problem 1.3 Show that if $n \geq k \geq r \geq s$, then

$$\binom{n}{k}\binom{k}{r}\binom{r}{s} = \binom{n}{s}\binom{n-s}{r-s}\binom{n-r}{k-r}.$$

Problem 1.4 A spider has 8 feet, 8 different shoes and 8 different socks. Find the number of ways in which the spider can put on the 8 socks and the 8 shoes (considering the order in which it puts them on). The only rule is that to put a shoe on the spider must already have a sock on that foot.

Problem 1.5 (OMCC 2003) A square board with side-length of 8 cm is divided into 64 squares with side-length of 1 cm each. Each square can be painted black or white. Find the total number of ways to color the board so that every square with side-length of 2 cm formed with 4 small squares with a common vertex has two black squares and two white squares.

Problem 1.6 (Colombia 2011) Ivan and Alexander write lists of integers. Ivan writes all the lists of length n with elements $a_1, a_2, \ldots, a_n$ such that $|a_1| + |a_2| + \cdots + |a_n| \leq k$. Alexander writes all the lists with length k with elements $b_1, b_2, \ldots, b_k$ such that $|b_1| + |b_2| + \cdots + |b_k| \leq n$. Prove that Alexander and Ivan wrote the same number of lists.

Problem 1.7 (China 2010) Let $A_1, A_2, \ldots, A_{2n}$ be pairwise different subsets of $\{1, 2, \ldots, n\}$. Determine the maximum value of

$$\sum_{i=1}^{2n} \frac{|A_i \cap A_{i+1}|}{|A_i| \cdot |A_{i+1}|},$$

with the convention that $A_{2n+1} = A_1$.

Problem 1.8 (Russia 2011) A table consists of n rows and 10 columns. Each cell of this table contains a digit (i.e. an integer from 0 to 9). Suppose one knows that for every row and every pair of columns B and C there exists a row that differs from A exactly in columns B and C. Prove that $n \geq 512$.

---

[4]In chess, a rook attacks all the pieces in its row and in its column.

Problem 1.9 (IMO 2000) A magician has cards numbered from 1 to 100, distributed in 3 boxes of different colors so that no box is empty. His trick consists in letting one person of the crowd choose two cards from different boxes without the magician watching. Then the person tells the magician the sum of the numbers on the two cards and he has to guess from which box no card was taken. In how many ways can the magician distribute the cards so that his trick always works?

Problem 1.10 Show that if n is a non-negative integer, then

$$\sum_{k=0}^{n} k \binom{n}{k}^2 = n \binom{2n-1}{n-1}.$$

Problem 1.11 (IMO 2003) Let A be a subset of 101 elements of $S = \{1, 2, \ldots, 1000000\}$. Show that there are numbers $t_1, t_2, \ldots, t_{100}$ in S such that the sets

$$A_j = \{x + t_j \mid x \in A\}, \quad j = 1, 2, \ldots, 100,$$

are pairwise disjoint.

Problem 1.12 Let m and r be non-negative integers such that $m > r$. Show that

$$\sum_{s=r}^{m} \binom{m}{s} \binom{s}{r} (-1)^{m-s} = 0.$$

Problem 1.13 (Japan 2009) Let N be a positive integer. Suppose that a collection of integers was written in a blackboard so that the following properties hold:

- each written number k satisfies $1 \le k \le N$;
- each k with $1 \le k \le N$ was written at least once;
- the sum of all written numbers is even.

Show that it is possible to label some numbers with $\circ$ and the rest with $\times$ so that the sum of all numbers with $\circ$ is the same as the sum of all numbers with $\times$.

Problem 1.14 (IMO 2006) We say that a diagonal of a regular polygon P of 2006 sides is a good segment if its extremities divide the boundary of P into two parts, each with an odd number of sides. The sides of P are also considered good segments. Suppose that P is divided into triangles using 2003 diagonals such that no two of them meet in the interior of P. Find the maximum number of isosceles triangles with two good segments as sides that can be in this triangulation.

The Pigeonhole Principle

2

2.1   The Pigeonhole Principle

The **pigeonhole principle** is one of the most used tools in combinatorics, and one of the simplest ones. It is applied frequently in graph theory, enumerative combinatorics and combinatorial geometry. Its applications reach other areas of mathematics, like number theory and analysis, among others. In olympiad combinatorics problems, using this principle is a golden rule and one must always be looking for a way to apply it. The first use of the pigeonhole principle is said to be by Dirichlet in 1834, and for this reason it is also known as the Dirichlet principle.

Proposition 2.1.1 (Pigeonhole principle) *If $n + 1$ objects are arranged in n places, there must be at least two objects in the same place.*

The proof of this proposition is almost immediate. If in every place there would be at most one object, we would have at most n objects, which contradicts the hypothesis. However, with the same reasoning we can prove a stronger version.

Proposition 2.1.2 (Pigeonhole principle, strong version) *If n objects are arranged in k places, there are at least $\lceil \frac{n}{k} \rceil$ objects in the same place.*

$\lceil x \rceil$ represents the smallest integer that is greater than or equal to x. Even though this result seems rather inconspicuous, it has very strong applications and is used frequently in olympiad problems.

Example 2.1.3 Given a triangle in the plane, prove that there is no line that does not go through any of its vertices but intersects all three sides.

Solution Any line divides the plane into two parts. By the pigeonhole principle, since there are three vertices, there must be at least two on the same side. The triangle side formed by those two vertices does not intersect the line.   □

P. Soberón, *Problem-Solving Methods in Combinatorics*,
DOI 10.1007/978-3-0348-0597-1_2, © Springer Basel 2013

Example 2.1.4 Prove that given 13 points with integer coordinates, one can always find 4 of them such that their center of gravity[1] has integer coordinates.

Solution Let us regard the coordinates modulo 2. There are only 4 possibilities: $(0,0)$, $(0,1)$, $(1,0)$ and $(1,1)$. So, by the pigeonhole principle, out of every 5 there must be two that have the same coordinates modulo 2. Let us take 2 such points and separate them from the others. We can continue this process and remove pairs with the same parity modulo 2 until we only have 3 points left. At that moment we have 5 different pairs which sum to 0 modulo 2 in both entries. Each sum modulo 4 can be 0 or 2, which gives only 4 possibilities for the sums modulo 4. Since we have 5 pairs, there must be 2 of them whose sums have the same entries modulo 4. These 4 points are the ones we were looking for.   □

Exercise 2.1.5 Find 12 points in the plane with integer coordinates such that the center of gravity of any 4 of them does not have integer coordinates.

Example 2.1.6 (Russia 2000) A 100×100 board is divided into unit squares. These squares are colored with 4 colors so that every row and every column has 25 squares of each color. Prove that there are 2 rows and 2 columns such that their 4 intersections are painted with different colors.

Solution Let us count the number P of pairs of squares (a_1, a_2) such that a_1 and a_2 are in the same row and have different colors. In every row there are 25 squares of each color, so there are $\binom{4}{2} 25 \cdot 25 = 6 \cdot 25 \cdot 25$ such pairs in each row. Thus $P = 100 \cdot 6 \cdot 25 \cdot 25$. We know there are $\binom{100}{2}$ pairs of columns and each of the P pairs must be in one of these pairs of columns. By the pigeonhole principle, there is a pair of columns that has at least

$$\frac{P}{\binom{100}{2}} = \frac{100 \cdot 6 \cdot 25 \cdot 25}{\frac{100 \cdot 99}{2}} = \frac{12 \cdot 25 \cdot 25}{99} = \frac{100 \cdot 75}{99} > 75$$

of the P pairs. From now on we only consider the pairs of squares in the same row, with different colors and that use these two columns.

    If these are not the columns we are looking for, then for any two of the pairs of squares we just mentioned there is at least one color they share. Take one of these pairs, and suppose it uses colors black and blue. Since there are more than 50 of these pairs, there is at least one that does not have color black. If it did not have

---

[1]The center of gravity of the set S of points (x_1, y_1), (x_2, y_2), $\ldots$, (x_n, y_n) in the plane is defined as the point

$$\left(\frac{x_1 + x_2 + \cdots + x_n}{n}, \frac{y_1 + y_2 + \cdots + y_n}{n} \right).$$

That is, it is the average of the set of points.

color blue, we would be done, so it must have blue and some other color (suppose it is green). In the same way there must be a pair that does not use color blue. Since it must share at least one color with the first pair, it must have black, and since it must share at least one color with the second pair it must have green. So the other pair is black and green.

Once we know we have these 3 pairs, any other pair must be either black and blue, blue and green or green and black. Since we have more than 75 of these pairs, these colors are used more than 150 times. However, each color is used only 25 times in each of the columns, so they can be used at most 150 times in total, which contradicts the previous statement. Thus, there are two rows of the kind we are looking for in the problem. □

A different version of the pigeonhole principle can also be used for infinite sets.

Proposition 2.1.7 (Infinite pigeonhole principle) *Given an infinite set of objects, if they are arranged in a finite number of places, there is at least one place with an infinite number of objects.*

The proof is analogous to the one for the pigeonhole principle: if in every place there is a finite number of objects, in total there would be a finite number of objects, which is not true.

Example 2.1.8 A 100×100 board is divided into unit squares. In every square there is an arrow that points up, down, left or right. The board square is surrounded by a wall, except for the right side of the top right corner square. An insect is placed in one of the squares. Each second, the insect moves one unit in the direction of the arrow in its square. When the insect moves, the arrow of the square it was in moves 90 degrees clockwise. If the indicated movement cannot be done, the insect does not move that second, but the arrow in its squares does move. Is it possible that the insect never leaves the board?

Solution We are going to prove that regardless of how the arrows are or where the insect is placed, it always leaves the board. Suppose this is not true, i.e., the insect is trapped. In this case, the insect makes an infinite number of steps in the board. Since there are only 100^2 squares, by the infinite pigeonhole principle, there is a square that is visited an infinite number of times.

Each time the insect goes through this square, the arrow in there moves. Thus, the insect was also an infinite number of times in each of the neighboring squares. By repeating this argument, the insect also visited an infinite number of times each of the neighbors of those squares. In this way we conclude that the insect visited an infinite number of times each square in the board, in particular the top right corner. This is impossible, because when that arrow points to the right the insect leaves the board. □

If one is careful enough, one can solve the previous example using only the finite version of the pigeonhole principle. However, doing it with the infinite version is much easier.

2.2   Ramsey Numbers

Example 2.2.1 Prove that in a party of 6 persons there are always three of them who know each other, or three of them such that no two of them know each other.

Solution Consider 6 points in the plane, one for each person in the party. We are going to draw a blue line segment between two points if the persons they represent know each other and a green line segment if they do not. We want to prove that there is either a blue triangle or a green triangle with its vertices in the original points.

Let v_0 be one of the points. From v_0 there are 5 lines going out to the other points, which are painted with two colors. By the pigeonhole principle, there are at least three of these lines painted in the same color (suppose it is blue). Let v_1, v_2, v_3 be three points that are connected with v_0 with a blue segment. If there is a blue segment between any two of them, those two vertices and v_0 form a blue triangle. If there is no blue segment between any two of them, then v_1, v_2, v_3 form a green triangle.   □

By solving this problem, we have proved that if we place 6 points in the plane and we join them with lines of two colors, there are three of them that form a triangle of only one color. The question now is if we can generalize this to bigger sets, when we are no longer looking for triangles of one color. That is, given two positive integers l and s, is there a number n large enough such that by placing n points in the plane and joining them with blue or green lines, there are always l of them such that all lines between two of them are blue or there are s of them such that all lines between two of them are green? In the previous example we saw that if $l = s = 3$, then $n = 6$ works. If there are such numbers n, we are interested in finding the smallest one that satisfies this property. If such number exists, it is denoted by

$$r(l, s)$$

and it is called the "**Ramsey's number** of (l, s)".

Exercise 2.2.2 Prove that if $l = s = 3$, then $n = 5$ is not enough.

Exercise 2.2.3 Let l be a positive integer. Prove that $r(2, l)$ exists and that $r(2, l) = l$.

Proposition 2.2.4 *For each pair (l, s) of positive integers the Ramsey number $r(l, s)$ exists, and if $l, s \geq 2$, then $r(l, s) \leq r(l - 1, s) + r(l, s - 1)$.*

Proof We prove this by induction on $l + s$. By Exercise 2.2.3, we know that if one of l, s is at most 2, $r(l, s)$ exists. This covers all cases with $l + s \leq 5$.

Suppose that if $l + s = k - 1$ then $r(l, s)$ exists. We want to prove that if $l + s = k$ then $r(l, s)$ exists. If any of l, s is at most 2, we have already done those cases, so we can suppose $l, s \geq 3$. Notice that $l + (s - 1) = (l - 1) + s = k - 1$, so $r(l, s - 1)$ and $r(l - 1, s)$ exist.

Fig. 2.1 p_0 is joined with
blue lines with A and with
green lines with B

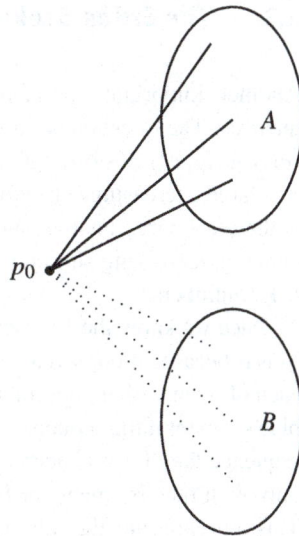

Consider a set of $r(l-1,s) + r(l,s-1)$ points in the plane joined with blue or green line segments. We want to prove that there are l of them joined only by blue segments or there are s of them joined only by green segments. Since $r(l,s)$ would be the minimum number that satisfies this, we would obtain that $r(l,s)$ exists and that $r(l,s) \le r(l-1,s) + r(l,s-1)$.

Let p_0 be one of the points. We claim that p_0 must be joined with at least $r(l-1,s)$ points with blue segments or at least with $r(l,s-1)$ points with green segments. If this does not happen, p_0 would be joined with at most $r(l-1,s)-1$ points with blue segments and at most with $r(l,s-1)-1$ points with green segments, so it would not be joined with the other $r(l-1,s)+r(l,s-1)-1$ points. (See Fig. 2.1.)

In other words, using the notation of the figure, we showed that either $|A| \ge r(l-1,s)$ or $|B| \ge r(l,s-1)$. Suppose that p_0 is joined with $r(l-1,s)$ points with blue segments. Among these points there are $l-1$ joined only with blue segments or s joined only with green segments. In the first case, these $l-1$ points and p_0 are only joined with blue segments, and in the second case, we have s points joined only with green segments. If p_0 is connected with $r(l,s-1)$ points with green segments, the way to find the sets is analogous. Thus $r(l,s)$ exists and $r(l,s) \le r(l-1,s) + r(l,s-1)$. $\square$

With this we have proven that the Ramsey numbers exist. However, finding specific Ramsey numbers turns out to be incredibly difficult. Up to this date no Ramsey number $r(l,s)$ is known with both $l,s \ge 5$.

2.3   The Erdős-Szekeres Theorem

Another important application of the pigeonhole principle is to number sequences. The question we are interested in is: If we have a long enough sequence
$(c_1, c_2, \ldots, c_k)$ of different numbers, when is there a large increasing subsequence
or a large decreasing subsequence? More precisely, given positive integers a and b,
is there a k large enough such that any sequence of k different numbers contains
either an increasing subsequence of $a + 1$ numbers or a decreasing subsequence of
$b + 1$ numbers?

Once we know the Ramsey numbers exist, it is easy to see that there is such a k.
This is because if our sequence has at least $r(a+1, b+1)$ numbers, we can associate
each of them with a point in the plane. Any two points are going to be joined with a
blue segment if the sequence they form is increasing and with a green segment if the
sequence they form is decreasing. Then, there must be $a + 1$ points that are joined
only with blue segments or $b + 1$ points that are joined only with green segments.
This set represents the subsequence we were looking for.

The bound we obtained in this process is $r(a + 1, b + 1)$, which is much larger
than we needed. The Erdős-Szekeres theorem gives us the best possible bound for
this result.

Theorem 2.3.1 (Erdős, Szekeres 1935) *Given any sequence of $ab + 1$ different numbers, there is always an increasing subsequence of at least $a + 1$ numbers or a
decreasing subsequence of at least $b + 1$ numbers.*

Proof Consider a sequence $(c_1, c_2, \ldots, c_{ab+1})$ of $ab + 1$ different numbers. To each
c_j of the sequence we assign a pair (a_j, b_j) of positive integers, where a_j is the
length of the longest increasing subsequence that ends in c_j and b_j is the length of
the longest decreasing subsequence that ends in c_j.

Given two numbers c_i and c_j in the sequence with $i < j$, we prove that their
pairs (a_i, b_i) and (a_j, b_j) cannot be equal. If $c_i < c_j$, we can add c_j to the largest
increasing subsequence that ends in c_i, so we have an increasing subsequence of
length $a_i + 1$ that ends in c_j. This gives $a_j \geq a_i + 1$. If $c_i > c_j$, we can add c_j to the
largest decreasing subsequence that ends in c_i, so we have a decreasing subsequence
of length $b_i + 1$ that ends in c_j. This gives $b_j \geq b_i + 1$.

If there were no subsequences of the lengths we were looking for, then for each
$1 \leq j \leq ab + 1$ we would have $a_j \leq a$ and $b_j \leq b$. This would give us at most ab
different pairs. Since we have $ab + 1$ pairs, by the pigeonhole principle at least two
must be equal, which is a contradiction. $\square$

Exercise 2.3.2 Given any two positive integers a and b, find a sequence of ab different real numbers with no increasing subsequences of length $a + 1$ or more and
no decreasing subsequences of length $b + 1$ or more.

2.4   An Application in Number Theory

Besides its importance in combinatorics, the pigeonhole principle has various strong applications in number theory. One of the most known ones is in showing that the decimal representation of any rational number[2] is periodic after some point. In other words, after some point is begins to repeat itself.

In this section we show a different application, establishing that every prime number of the form $4k + 1$ can be written as the sum of two squares. To see this we need the following proposition:

Proposition 2.4.1 *For any integers n and u, there are integers x and y not both 0 such that $-\sqrt{n} \leq x \leq \sqrt{n}$, $-\sqrt{n} \leq y \leq \sqrt{n}$ and $x - uy$ is divisible by n.*

Proof Let $k + 1 = \lfloor \sqrt{n} \rfloor$ be the largest integer that is smaller than or equal to $\sqrt{n}$, that is, $k \leq \sqrt{n} < k + 1$. Consider the numbers of the form $x - uy$ with x and y in $\{0, 1, 2, \ldots, k\}$. Each has $k + 1$ options, so there are $(k + 1)^2 > n$ possible numbers. Thus (by the pigeonhole principle!), there are two that leave the same remainder when divided by n. If they are $x_1 - uy_1$ and $x_2 - uy_2$, their difference $(x_1 - x_2) - u(y_1 - y_2)$ is divisible by n. If we take $x = x_1 - x_2$ and $y = y_1 - y_2$, we have that they are not both 0, since the pairs (x_1, y_1) and (x_2, y_2) were different. Also, x and y are in the desired intervals. □

With this we are ready to prove that every prime number of the form $4k + 1$ is the sum of two squares. The only thing we need to know about these numbers is that for every prime p of the form $4k + 1$ there is a u such that $u^2 + 1$ is divisible by p. This last result can also be proven in a combinatorial way, counting how many pairs of numbers a, b in $\{0, 1, 2, \ldots, p - 1\}$ satisfy $ab \equiv -1 \pmod{p}$ and how many pairs of different numbers a, b satisfy $ab \equiv 1 \pmod{p}$. However, we will not do it in this book.

Theorem 2.4.2 (Fermat) *Every prime p of the form $4k + 1$ can be written as the sum of two squares.*

Proof Let u be an integer such that $u^2 + 1$ is divisible by p. Using Proposition 2.4.1 we know that there are integers x, y not both 0 such that $x - uy$ is divisible by p and $-\sqrt{p} \leq x \leq \sqrt{p}$, $-\sqrt{p} \leq y \leq \sqrt{p}$. The condition can be translated to $x^2 \leq p$ and $y^2 \leq p$. Since p is prime, it is not a perfect square, so the inequalities are strict. Since $x \equiv uy \pmod{p}$, we have that $x^2 \equiv u^2 y^2 \equiv -y^2 \pmod{p}$, so $x^2 + y^2$ is divisible by p. However,

$$0 < x^2 + y^2 < 2p.$$

With this we have that $x^2 + y^2 = p$. □

---

[2] A rational number is one that can be written as the quotient of two integers.

It is also known that a prime p of the form $4k + 1$ can be written as the sum of two squares in a unique way, so this application of the pigeonhole principle finds the only pair that satisfies this. This shows that even though the technique seems elementary, it gives very precise results.

2.5   Problems

Problem 2.1 Show that given 13 points in the plane with integer coordinates, there are three of them whose center of gravity has integer coordinates.

Problem 2.2 Show that in a party there are always two persons who have shaken hands with the same number of persons.

Problem 2.3 (OIM 1998) In a meeting there are representatives of n countries ($n \geq 2$) sitting at a round table. It is known that for any two representatives of the same country their neighbors to their right cannot belong to the same country. Find the largest possible number of representatives in the meeting.

Problem 2.4 (OMM 2003) There are n boys and n girls in a party. Each boy likes a girls and each girl likes b boys. Find all pairs (a, b) such that there must always be a boy and a girl that like each other.

Problem 2.5 For each n show that there is a Fibonacci[3] number that ends in at least n zeros.

Problem 2.6 Show that given a subset of $n + 1$ elements of $\{1, 2, 3, \ldots, 2n\}$, there are two elements in that subset such that one is divisible by the other.

Problem 2.7 (Vietnam 2007) Given a regular 2007-gon, find the smallest positive integer k such that among any k vertices of the polygon there are 4 with the property that the convex quadrilateral they form shares 3 sides with the polygon.

Problem 2.8 (Cono Sur Olympiad 2007) Consider a 2007×2007 board. Some squares of the board are painted. The board is called "charrúa" if no row is completely painted and no column is completely painted.

- What is the maximum number k of painted squares in a charrúa board?
- For such k, find the number of different charrúa boards.

Problem 2.9 Show that if 6 points are placed in the plane and they are joined with blue or green segments, then at least two monochromatic triangles are formed with vertices in the 6 points.

---

[3]Fibonacci numbers are defined by the formulas (6.3) and (6.4).

Problem 2.10 (OMM 1998) The sides and diagonals of a regular octagon are colored black or red. Show that there are at least 7 monochromatic triangles with vertices in the vertices of the octagon.

Problem 2.11 (IMO 1964) 17 people communicate by mail with each other. In all their letters they only discuss one of three possible topics. Each pair of persons discusses only one topic. Show that there are at least three persons that discussed only one topic.

Problem 2.12 Show that if l, s are positive integers, then

$$r(l, s) \le \binom{l + s - 2}{l - 1}.$$

Problem 2.13 Show that $r(3, 4) = 9$.

Problem 2.14 Show that if an infinite number of points in the plane are joined with blue or green segments, there is always an infinite number of those points such that all the segments joining them are of only one color.

Problem 2.15 (Peru 2009) In the congress, three disjoint committees of 100 congressmen each are formed. Every pair of congressmen may know each other or not. Show that there are two congressmen from different committees such that in the third committee there are 17 congressmen that know both of them or there are 17 congressmen that know neither of them.

Problem 2.16 (IMO 1985) We are given 1985 positive integers such that none has a prime divisor greater than 23. Show that there are 4 of them whose product is the fourth power of an integer.

Problem 2.17 (Russia 1972) Show that if we are given 50 segments in a line, then there are 8 of them which are pairwise disjoint or 8 of them with a common point.

Problem 2.18 (IMO 1972) Show that given 10 positive integers of two digits each, there are two disjoint subsets A and B with the same sum of elements.

Problem 2.19 There are two circles of length 420. On one 420 points are marked and on the other some arcs of circumference are painted red such that their total length adds up less than 1. Show that there is a way to place one of the circles on top of the other so that no marked point is on a colored arc.

Problem 2.20 (Romania 2004) Let $n \ge 2$ be an integer and X a set of n elements. Let $A_1, A_2, \ldots, A_{101}$ be subsets of X such that the union of any 50 of them has more than $\frac{50n}{51}$ elements. Show that there are three of the A_j's such that the intersection of any two is not empty.

Problem 2.21 (Tournament of towns 1985) A class of 32 students is organized in 33 teams. Every team consists of three students and there are no identical teams. Show that there are two teams with exactly one common student.

Problem 2.22 (Great Britain 2011) Let G be the set of points (x, y) in the plane such that x and y are integers in the range $1 \leq x, y \leq 2011$. A subset S of G is said to be *parallelogram-free* if there is no proper parallelogram with all its vertices in S. Determine the largest possible size of a parallelogram-free subset of G.

Note: A proper parallelogram is one whose vertices do not all lie on the same line.

Problem 2.23 (Italy 2009) Let n be a positive integer. We say that k is n-square if for every coloring of the squares of a $2n \times k$ board with n colors there are two rows and two columns such that the 4 intersections they make are of the same color. Find the minimum k that is n-square.

Problem 2.24 (Romania 2009) Let n be a positive integer. A board of size $N = n^2 + 1$ is divided into unit squares with N rows and N columns. The N^2 squares are colored with one of N colors in such a way that each color was used N times. Show that, regardless of the coloring, there is a row or a column with at least $n + 1$ different colors.

Invariants

3

3.1 Definition and First Examples

Example 3.1.1 There are three piles with n tokens each. In every step we are allowed to choose two piles, take one token from each of those two piles and add a token to the third pile. Using these moves, is it possible to end up having only one token?

Solution To the tokens in the first pile we can assign the pair $(0, 1)$, to the tokens in the second pile the pair $(1, 0)$ and to the tokens in the third pile the pair $(1, 1)$. Notice that the sum modulo 2 of any two of these pairs give us the third one. Thus, in every step the sum modulo 2 of all the assigned pairs is the same. However, the sum of all the assigned pairs in the beginning is $(2n, 2n)$, which is equal to $(0, 0)$ modulo 2. Since this pair was not assigned to any pile, it is not possible to end up with only one token. □

In the previous example the strategy was to find a property that did not change in every step of the problem, and proving it could not be preserved in the end. When one deals with problems involving a transformation (such as taking off tokens in a certain way), a property that does not change under that transformation is called an **invariant**. Invariants can be extremely diverse, and there are numerous problems of international mathematical olympiads that can be solved by finding some very special invariant. When one is trying to solve olympiad problems (or any problem in mathematics), looking for invariants is a fundamental strategy.

Example 3.1.2 Let n be a positive integer and consider the ordered list $(1, 2, 3, \ldots, n)$. In each step we are allowed to take two different numbers in the list and swap them. Is it possible to obtain the original list after exactly 2009 steps?

Solution Suppose at some moment number 1 is in place a_1, number 2 is in place a_2, and so on. We are going to count the number T of pairs (x, y) such that $x < y$ but $a_x > a_y$. In other words, we are counting the pairs of numbers that are not ordered. Suppose that in a step we swapped numbers a and b, and there were k numbers

P. Soberón, *Problem-Solving Methods in Combinatorics*,
DOI 10.1007/978-3-0348-0597-1_3, © Springer Basel 2013

between them. If the pair (a, b) was counted in T, now it is not and vice-versa. The same happens to the k pairs formed by a and any of the numbers between a and b and the k pairs of b with any of those numbers. So there are exactly $2k + 1$ pairs changing from being counted in T or not being counted in T. This means that T changes by an odd number. Thus, after 2009 steps T must be odd. Since T cannot be 0 after 2009 steps, we cannot get the original order.   □

These ideas are useful even in combinatorial geometry problems.

Example 3.1.3 (IMO 2011) Let S be a finite set of at least two points in the plane. Assume that no three points of S are collinear. A *windmill* is a process that starts with a line l going through a single point $P \in S$. The line rotates clockwise about the *pivot* P until the first time that the line meets some other point belonging to S. This point, Q, takes over as the new pivot, and the line now rotates clockwise about Q, until it meets a point of S. This process continues indefinitely. Show that we can choose a point P in S and a line l going through P such that the resulting windmill uses each point of S infinitely many times.

Solution Consider the set of all lines that go through two points of S. Choose a direction l that is not parallel to any of those lines. Then, we can find a line l_0 parallel to l that goes through a point $P \in S$ and such that the absolute difference of the number of points on the two sides of l_0 is at most 1. That is, the closest we can get to dividing the points of S in half subject to passing through a point of S. Consider one of the half-planes determined by a line as the red half-plane and the other as the blue half-plane. Start the windmill with l_0 until it makes half of a complete turn. At this point consider its position l_1. l_1 is parallel to l_0, and the positions of the red and blue half-planes have changed. Note that the absolute difference between the number of points on the two sides of l_1 is also at most 1. Thus there are no points of S between l_0 and l_1. This means that, other than the pivots of l_0 and l_1, every point that was on the red half-plane is now on the blue half-plane and vice versa. Thus, the windmill used each point of S at least once. Since it will continue to use each point at least once per half turn, we are done.

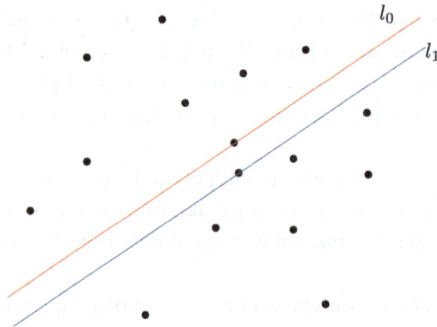

Figure showing l_0 and l_1 for a set of 16 points. If S has an odd number of points, then $l_0 = l_1$.   □

In the previous example the invariant was the absolute difference of the number of points on the two sides of the line (when it is not passing through 2 points of S). Sometimes it is not easy to find a property that does not change, but finding one that always changes the same way can be equally useful.

Example 3.1.4 23 friends want to play soccer. For this they choose a referee and the others split into two teams of 11 persons each. They want to do this so that the total weight of each team is the same. We know that they all have integer weights and that, regardless of who is the referee, it is possible to make the two teams. Prove that they all have the same weight.

Solution Let $a_1, a_2, \ldots, a_{23}$ denote the weights of the persons. Let $S = a_1 + a_2 + \cdots + a_{23}$. The condition tells us that if we remove the person with weight a_i, the rest can be split into two teams of weight X. Thus $S - a_i = 2X$. This tells us that a_i and S have the same parity. Since this can be done for any a_i, they are all even or all odd. We are going to generate another list $b_1, b_2, \ldots, b_{23}$ of new weights that still satisfies the conditions of the problem. If the a_i are even, we replace each one by $b_i = \frac{a_i}{2}$. If they are odd, we replace each one by $b_i = a_i - 1$.

It is clear that the new list of weights satisfies the conditions of the original problem, and that $b_i \geq 0$ for all i. If the a_i were not all 0, then the sum of the weights b_i is strictly smaller. If we keep repeating this step we are always reducing the sum of the weights, so eventually we reach a list of only zeros. Since we are able to do this, we conclude the numbers in the original list were all equal.   □

In the previous example we exhibited a way to modify the problem so that the sum S was always decreasing. The reason why we can finish is because there is no infinite sequence of non-negative integers that is strictly decreasing. This technique is known as **infinite descent**.

The question of what happens if the weights are not integers is also interesting. If all the friends have rational weights, multiplying all by some appropriate integer reduces the problem to the case with integer weights. If the weights are positive real numbers, then one needs to use linear algebra to reduce the problem to the rational case (simple approximation arguments do not work). Since this technique is beyond the purpose of this book, we will not show the details. However, the reader familiar with linear algebra should be able to deduce a solution with the previous comments.

Let us see how we can apply infinite descent to a much more difficult problem.

Example 3.1.5 (IMO 1986) We assign an integer to each vertex of a regular pentagon, so that the sum of all is positive. If three consecutive vertices have assigned numbers x, y, z, respectively, and $y < 0$, we are allowed to change the numbers (x, y, z) to $(x + y, -y, z + y)$. This transformation is made as long as one of the numbers is negative. Decide if this process always comes to an end.

Solution Let the numbers in the pentagon be a, b, c, d, e, as shown in the figure.

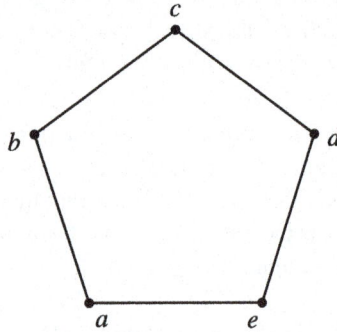

When one does the transformation, the sum of all numbers remains the same. Consider $S(a, b, c, d, e) = (a - c)^2 + (c - e)^2 + (e - b)^2 + (b - d)^2 + (d - a)^2$. It is clear that $S(a, b, c, d, e) \geq 0$. Suppose that c is negative and we do the change with c as the middle number. We have that

$$
\begin{aligned}
S(a, &b + c, -c, d + c, e) \\
&= (a + c)^2 + (-c - e)^2 + (e - b - c)^2 + (b - d)^2 + (d + c - a)^2 \\
&= \left[(a - c)^2 + 4ac \right] + \left[(c - e)^2 + 4ce \right] + \left[(e - b)^2 - 2ec + 2bc + c^2 \right] \\
&\quad + (b - d)^2 + \left[(d - a)^2 - 2ac + 2cd + c^2 \right] \\
&= S(a, b, c, d, e) + 2c(a + b + c + d + e).
\end{aligned}
$$

Since $c < 0$ and we assumed that $a + b + c + d + e > 0$, we have that $2c(a + b + c + d + e) < 0$. Therefore, $S(a, b, c, d, e)$ is reduced after every transformation. By infinite descent, there must be a moment when we can no longer do the transformation, which is what we wanted to prove.   □

It should be noted that this problem was solved completely by a very small number of students (as it should be expected for a Problem 3 at the IMO). The invariant S here might seem at first quite unnatural, but that is not the case. The first thing a student has to note is that the sum of all the numbers is constant and positive, and thus so is their average. This means that if every number was as close as possible to the average, then none would be negative. This makes it natural to search for functions that increase as the numbers are different from their average, or different from each other. Given two numbers x and y, $(x - y)^2$ or $|x - y|$ are good ways to see how different they are. If we start with $(x - y)^2$ and try to apply it to 5 numbers, then the function S (or one of the same kind) arises naturally. Using absolute values it is also possible to obtain a function that behaves like S.

It should be noted that this type of ideas are very useful even outside olympiad problems. Given a set of number $a_1, a_2, \ldots, a_n$ with average λ, the number $\mu_2 = \sum_{i=1}^{n} (a_i - \lambda)^2$ is called the **second moment** of those numbers. This number and its generalizations are widely used in probability theory and its applications, but we leave further investigations to the interested reader.

3.2   Colorings

Example 3.2.1 Is it possible to cover a 10×10 board with the following pieces without them overlapping?

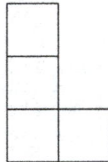

Note: The pieces can be flipped and turned.

Solution Color the columns white and black alternatingly. There are 50 white squares and 50 black squares. We can see that, regardless of how we place a piece, it always covers three squares of one color and one of the other. Let us call the piece black if it covers three black squares and white if it covers three white squares. Then the number of black pieces is equal to the number of white pieces. This tells us that the total number of pieces must be even. This would mean that the number of squares should be divisible by 8. Since there are 100 squares, there is no possible cover.   □

In the previous example the key to see that we needed an even number of pieces was to color the columns. Coloring is a very neat technique in problems involving boards since it allows us to simplify the problem a great deal. The important part is focusing on an adequate subset of the squares (in the above example the subset consisted of the squares in even columns), however doing it with colors is a lot easier.

The kinds of colorings can be very different. In the previous problem we used only two colors, but sometimes more are needed. There is no general rule for determining which one is going to solve the problem. There are some colorings (such as a chessboard coloring) that are frequently used, but the only way to learn how to use this technique is by solving several problems of this style. The problems in international contests that can be solved by coloring usually have the property that commonly used colorings do not give much information. They involve specific colorings for that problem.

When the problem is related to pieces covering a certain figure, the "good colorings" are those that yield an invariant associated with the pieces. This can be the number of squares of one color they cover, the number of colors they may use, some parity argument, etc. Coloring is basically an illustrative way to describe invariants.

Example 3.2.2 On a 9×9 board 65 insects are placed in the centers of some of the squares. The insects start moving at the same time and speed to a square that shares a side with the one they were in. When they reach the center of that square, they make a 90 degrees turn and keep walking (without leaving the board). Prove that at some moment of time there are two insects in the same square.

Note: When they turn it can be either to the right or to the left.

Solution Let us color the board using 4 colors A, B, C and D the following way:

A	C	A	C	A
B	D	B	D	B
A	C	A	C	A	...
B	D	B	D	B
A	C	A	C	A
		⋮		⋱

There are 25 A squares, 20 C squares, 20 B squares and 16 D squares. By the pigeonhole principle, there are at least 33 insect in colors A and D or 33 insects in colors B and C. In the second case, after one step there are at least 33 insect in squares A and D. Since the insects have to turn after every step, we know that every insect that was in an A square goes to a D square after two steps and vice-versa. By the pigeonhole principle there must be at least 17 insects in A squares or at least 17 insect in D squares. This tells us that at some moment there are at least 17 insects in D squares. Again, by the pigeonhole principle, there are two insects in the same square. □

Example 3.2.3 (IMO shortlist 2002) A $(2n - 1) \times (2n - 1)$ board is going to be tiled with pieces of the type as shown in Fig. 3.1.

Prove that at least $4n - 1$ of the first type will be used.

Solution Number the rows and columns from 1 to $2n - 1$. Color in black the squares that are in an odd row and an odd column and color white the rest of the squares. There are n^2 black squares and $3n^2 - 4n + 1$ white squares. The pieces of the first type can cover two white squares and one black square or three white squares. The rest of the pieces cover always three white squares and one black square. Let A be the number of pieces of the first kind that cover one black square, B the number of pieces of the first kind that cover no black squares and C the number of pieces of the other kinds.

Fig. 3.1 Pieces for Example 3.2.3

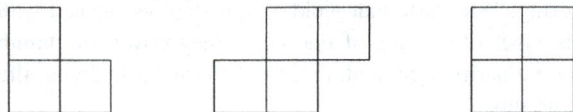

Counting the number of black squares we have that $A + C = n^2$ and counting the white squares we have that $3n^2 - 4n + 1 = 2A + 3B + 3C$. So $4n - 1 = 3n^2 - (3n^2 - 4n + 1) = 3(A + C) - (2A + 3B + 3C) = A - B \le A + B$. Thus there are at least $4n - 1$ pieces of the first kind. □

In the previous example a very special coloring was used. It is interesting to note that a chessboard coloring provides almost no information, and the coloring of the first example of the section only shows that at least $2n - 1$ pieces of the first kind are necessary.

Another difficulty of this problem is that finding a cover of an odd square board with this pieces is not easy. In fact, for $n = 1, 2, 3$ the number of squares needed by the pieces of the first kind is greater than the number of squares in the board, so no tiling is possible. In the following figure we show how to tile a 7×7 board (when $n = 4$).

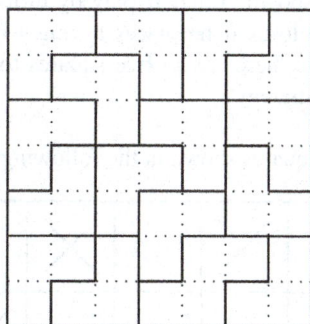

3.3 Problems Involving Games

Example 3.3.1 (OMCC 2001) A and B are going to play a game by turns. Before they start, they form a circle with 2001 other persons. At every turn they can remove one of their neighbors from the circle. The winner is the one who gets the other person out of the circle. If A starts, decide who has a winning strategy.

Note: The other 2001 persons do not have turns.

Solution When the game starts there are 2001 other persons. This means that A and B divide the circle in two arcs, one of which has an odd number of persons and the other an even number. The strategy for A is to remove always a person on the even side. This leaves B with an odd number of persons on each side. When B plays, A has again a side with an odd number of persons and an even number on the other, so he can continue with his strategy. B can never hope to win if A plays this way, since he always has at least one person between himself and A on both sides. Since the game must end after at most 2001 turns, A wins. □

In problems involving games, finding a winning strategy means finding a way to play so that, regardless of how the other person plays, one is going to win. The key to solve this kind of problems is to find an invariant in the game and exploit it. The invariant has to be a certain state of the game. We are looking for a state with the following properties:

- A person in that state cannot win the game.
- If the other person played in that state, we can force him back to that state.

This kind of state is called a **losing position**. The positions that can send the other player to a losing position are called **winning positions**.

In the previous example the losing position was having an odd number of persons on each side. In this type of problems, to find the losing positions it is convenient to look at the positions near the end of the game when one cannot win and work backwards in the game. It is also a good strategy to try a few games to look for the invariant.

Example 3.3.2 On a chessboard A and B play by turns to place black and white knights, respectively. One loses if he places a knight on a square attacked by a knight of the other color or there are no free squares to place the knight on. If A starts, who has a winning strategy?

A knight attacks the 8 squares shown in the following figure.

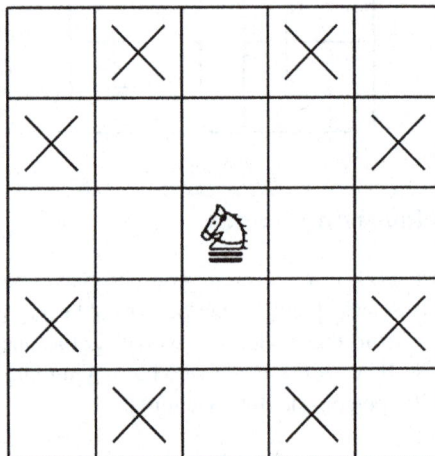

Solution We will prove that B has a winning strategy. The strategy is to do the same thing as A but symmetrically to the center of the board. If B plays this way, A always plays on a board that is symmetric with respect to its the center. If A could place a knight without being attacked by B's knights, this means that on the square symmetric with respect to the center B can place a knight without being attacked by A's knights. Given the way knights attack, A can never place a knight that attacks the square where B would play. Thus A loses in the end.   □

Exercise 3.3.3 Solve the previous problem on a 9×9 chessboard.

Example 3.3.4 There are 2010 matches on a table. A and B play by turns to remove matches from the table. At each turn, they must remove $1, 3, 4, 5$ or 7 matches. Whoever removes the last match wins. If A plays first, who has a winning strategy?

Solution Notice that the multiples of 8 are losing positions. If a player is in a multiple of 8, he cannot take all the matches. If he takes a matches, the other player can take $8 - a$ and bring him back to a multiple of 8. Since 2010 leaves remainder 2 when divided by 8, if the first player does not remove 4 matches, the other player can leave him in a multiple of 8. The same happens for the second player. If they both play only removing 4 matches each, B leaves 2 matches in his last turn, so B has the winning strategy. □

Example 3.3.5 (Bulgaria 2001) A and B play by turns to write ones and zeros in a list, from left to right. The game ends when each has written 2001 numbers. When the game ends the sequence of numbers is seen as the expansion of a number in base 2. A wins if that number can be written as the sum of two perfect squares and B wins otherwise. Prove that B has a winning strategy.

Solution We consider the following strategy for B: If at any moment A writes a 1, then B only copies A's moves. The reason for this is that by following this strategy, we end up with a number with an even number of zeros at the end. We know that $4m$ can be written as the sum of two squares if and only if m can be written as the sum of two squares. Thus we can ignore the zeros at the end and suppose that the number finishes with 11, which means that the number is congruent to 3 modulo 4. From this, we infer that the number cannot be written as the sum of two squares. If A only writes 0, then B writes 1998 times 0 and then 3 times 1. This gives at the end the binary expansion of the number 21, which cannot be written as the sum of two squares. □

The crucial step in this problem was to find a property with which a number could not be written as the sum of two squares. This was precisely being 3 modulo 4. Seeing how this property translates to the game gives the winning strategy if A plays a 1. In the other case, since we know how A is going to play it suffices to try a few cases.

This kind of problems belong to of a branch of mathematics called game theory. In that branch you work with situations where one or more persons have to make some kind of decision to maximize a profit. This kind of situations are called games. In the games we have seen the profit was absolute, namely, it was "winning" the game. The objective of game theory is to find strategies to maximize such profit, and analyzing the situation when everyone plays with an optimal strategy (these situations are called **equilibriums**). Let us see another (very simple) example where the objective is no longer simply "winning" the game.

Example 3.3.6 A pirate ship has 2009 treasure chests (all chests are closed). Each chest contains some amount of gold. To distribute the gold the pirates are going to do the following. The captain is going to decide first how many chests he wants to keep and tell that number to the rest of the pirates. Then he is going to open all the chests and decide which ones he wants to keep (he can only choose as many as he said before opening them). The captain wants to make sure he can keep at least half of the total gold. However, he wants to say the smallest possible number to keep the rest of the pirates as happy as he can. What number should the captain say?

Note: The chests may be empty.

Solution Note that the chests could all contain the same amount of gold, so to keep at least of half of the gold the captain should keep at least half of the chests. With this in mind, the number he must say is at least 1005. Let us see that regardless of the gold distribution, 1005 of the chests are enough to keep half of the gold. To do this it is enough to open the chests and order them according to the amount of gold they have $a_1 \geq a_2 \geq a_3 \geq \cdots \geq a_{2009}$. If the captain keeps the chests $a_1, a_2, \ldots, a_{1005}$ then he has at least half of the gold. $\qquad\square$

The previous example becomes interesting if further complications are added.

Example 3.3.7 A pirate ship has 2009 treasure chests (all chests are closed). Each chest contains some amount of gold and some amount of silver. To distribute the gold and silver the pirates are going to do the following. The captain is going to decide first how many chests he wants to keep and tell that number to the rest of the pirates. Then he is going to open all the chests and decide which ones he wants to keep. The captain wants to make sure he can keep at least half of the total gold and half of the total silver. However, he wants to say the smallest possible number to keep the rest of the pirates as happy as he can. What number should the captain say?

Note: The amount of gold and silver in each chest may be different.

Solution Since the captain wants to keep at least half of the gold, he needs again at least 1005 chests. It turns out that 1005 is also enough in this case. To see this, let us order again the chests according to their amount of gold $a_1 \geq a_2 \geq \cdots \geq a_{2009}$. The choice of the captain is as follows. First he splits the chests into the following groups: $\{a_1\}, \{a_2, a_3\}, \{a_4, a_5\}, \ldots, \{a_{2008}, a_{2009}\}$. From each group he picks the one that has more silver. By doing this he guarantees he keeps half of the total amount of silver. However, in each group the chest he did not choose has at most the same amount of gold as the chest he chose from the previous group. Thus he also has at least half of the total amount of gold. $\qquad\square$

Game theory was created to abstract and solve problems in economics. However, due to the nature of the objects it studies, it has a wide number of applications. It is closely related to other fields of mathematics, such as probability. However, it also has applications to other subjects like political sciences, biology, philosophy and even military strategies.

3.4   Problems

Problem 3.1 On an $m \times n$ board a path is a sequence of squares such that any two consecutive squares share one side. Show that on an $m \times n$ board there is a path that starts and ends in the same square and goes through every other square exactly once if and only if at least one of m, n is even and both are greater than or equal to two.

Note: In this problem, if $m = n = 1$, we do not consider the single square as a valid path.

Problem 3.2 On a table there are 2009 tokens which are red on one side and black on the other. A and B play by turns. In each turn it is permitted to remove any number of tokens from one color or flip any number of tokens of the same color. The one that removes the last token wins. If A plays first, who has a winning strategy?

Problem 3.3 (OMCC 2003) A and B play with a set of 2003 coins by turns. In each turn it is permitted to remove a number of coins that is a divisor of the number of coins remaining. The one that removes the last coin loses. If A plays first, who has a winning strategy?

Problem 3.4 (OIM shortlist 2009) On an 8×8 board there is a lamp in every square. Initially every lamp is turned off. In a move we choose a lamp and a direction (it can be the vertical direction or the horizontal one) and change the state of that lamp and all its neighbors in that direction. After a certain number of moves, there is exactly one lamp turned on. Find all the possible positions of that lamp.

Problem 3.5 (OIM 2004) We are given a 1001×1001 board. We want to color some squares so that the following two conditions are met:

- If two squares share a side, at least one of them is colored.
- If 6 squares are consecutive (horizontally or vertically), then among them at least two consecutive squares are colored.

Find the minimum number of squares that must be colored under these two rules.

Problem 3.6 (Germany 2009) On a table there are 100 coins. A and B are going to remove coins from the table by turns. In each turn they can remove 2, 5 or 6 coins. The first one that cannot make a move loses. Determine who has a winning strategy if A plays first.

Problem 3.7 (Middle European Mathematical Olympiad 2010) All positive divisors of a positive integer N are written on a blackboard. Two players A and B play the following game, taking alternate moves. In the first move, the player A erases N. If the last erased number is d, then the next player erases either a divisor of d or a multiple of d. The player who cannot make a move loses. Determine all numbers N for which A has a winning strategy.

Problem 3.8 In a 4×4 board the numbers from 1 to 15 are arranged in the following way:

1	2	3	4
5	6	7	8
9	10	11	12
13	15	14	

In a move we can move some number that is in a square sharing a side with the empty square to that square. Is it possible to reach the following position using these moves?

1	2	3	4
5	6	7	8
9	10	11	12
13	14	15	

Problem 3.9 (OIM 2000) On a table there are 2000 coins. A and B are going to play by turns to remove coins from the table. In each turn they may remove 1, 2, 3, 4 or 5 coins from the table but they cannot remove the same number as the other person removed in the previous turn. The one that removes the last coin wins. If A plays first, who has a winning strategy?

Problem 3.10 (OMM 2003) In a box we have cards with all the pairs (a, b) with $1 \leq a < b \leq 2003$ (no cards are repeated and all cards have only one pair). A and B are going to play by turns to remove cards from the box. At each turn, when they remove a card with the pair (a, b), they write on a blackboard the product ab of the numbers in the card. The first one to make the greatest common divisor of the numbers on the blackboard to be 1 loses. If A plays first, who has a winning strategy?

Problem 3.11 (APMO 2007) In each square of a 5×5 board there is a lamp turned off. If we touch a lamp then that lamp and the ones in neighboring squares change their states. After a certain number of moves there is exactly one lamp turned on. Find all squares in which that lamp can be.

Note: Neighboring squares are squares that share a side.

Problem 3.12 (Italy 2007) On a table there are cards with the numbers $0, 1, 2, \ldots,$ 1024. A and B play by turns to remove cards from the table. First B removes 2^9 cards. Then A removes 2^8 cards. Then B removes 2^7 cards and so on until there are exactly 2 cards left with numbers a and b. A has to pay $|a - b|$ dollars to B. What is the largest amount of money that B can guarantee he will win?

Problem 3.13 (Japan 2011) Let N be a positive integer. N squares are lined up contiguously from left to right. Students A and B play a game according to the following rules:

- To start off, A will write one non-negative integer into each of the N squares.
- The game ends when the following condition is achieved: for every i satisfying $1 \leq i \leq N - 1$, the number in the i-th square from the left is less than or equal to the number in the $(i + 1)$-th square from the left.

The game continues as long as this condition is not achieved by repeating the following procedure:

(a) A designates one non-negative integer.
(b) B chooses one of the N squares and replaces the number in that square by the number designated by A in (a).

Is it possible for B to force the game to end regardless of how A plays?

Problem 3.14 (USAMO 1994) We are given a regular 99-gon with sides painted in red, blue or yellow. Initially they are painted in the following way: red, blue, red, blue, $\ldots$, red, blue, yellow. In a step we can change the color of any side as long as we do not generate two adjacent sides of the same color. Is it possible to reach a coloring red, blue, red, blue, $\ldots$, red, yellow, blue after some finite number of steps?

    Note: The two colorings are considered in clockwise order.

Problem 3.15 (Romania 2007) In an $n \times n$ board the squares are painted black or white. Three of the squares in the corners are white and one is black. Show that there is a 2×2 square with an odd number of white unit squares.

Problem 3.16 (Argentina 2009) Consider an $a \times b$ board, with a and b integers greater than or equal to two. Initially all the squares are painted white and black as a chessboard. The allowed operation is to choose two unit squares that share one side and recolor them in the following way: Any white square is painted black, any black square is painted green and any green square is painted white. Determine for which values of a and b it is possible, using this operation several times, to get all the original black squares to be painted white and all the original white squares to be painted black.

    Note: Initially there are no green squares, but these appear after the first time we use the operation.

Problem 3.17 (Italy 2008) *A* and *B* are going to play by turns with piles of coins on a table (there are no piles with 0 coins). In each turn they can do one of the following moves:

- Choose a pile with an even number of coins and split it into two piles with the same number of coins each.
- Remove from the table all the piles with an odd number of coins. There has to be at least one odd pile to do this.

In each turn, if one of these two moves is impossible, they have to perform the other. The first one that cannot make a valid move loses. If *A* plays first and initially there is only one pile with 2008^{2008} coins, who has a winning strategy? What are the initial conditions under which *A* has a winning strategy?

Problem 3.18 (Vietnam 1993) With the following shapes we tile a 1993×2000 board. Let *s* be the number of shapes used of the first two types. Find the largest possible value of *s*.

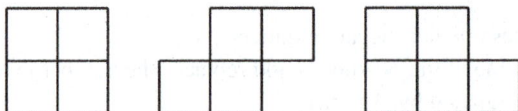

Problem 3.19 We are given a 5×5 board with a chessboard coloring where the corners are black. In each black square there is a black token and in each white square there is a white token. The tokens can move to neighbor squares (squares that share a side with the one they are on). *A* and *B* are going to play by turns in the following way: First *A* chooses a black token and removes it from the board. Then, *A* moves a white token to the empty space. Then *B* moves a black token to the empty space. At each of his following turns, *A* moves a white token to the empty space and *B* moves a black token to the empty space. The game ends when one of them cannot make a valid move, and that person loses. Is there a winning strategy? If so, who has it?

Problem 3.20 (OMM 1999) We say a polygon is orthogonal if all its angles are of 90 or 270 degrees. Show that an orthogonal polygon whose sides are all of odd integer length cannot be tiled with 2×1 domino tiles.

   Note: The polygons cannot have holes. (See Fig. 3.2.)

Fig. 3.2 Example of an
orthogonal polygon

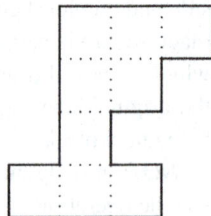

Problem 3.21 (USAMO 1999) On a 1×2000 board, A and B play by turns to write S or O in each square. The first one to write the word SOS in three consecutive squares wins. If A plays first, show that B has a winning strategy.

Problem 3.22 (IMO 2004) Find all pairs (m, n) such that an $m \times n$ board can be tiled with the following tile:

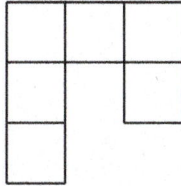

Note: The tile can be rotated and flipped upside down.

Problem 3.23 (IMO shortlist 1999) Show that if a $5 \times n$ board can be tiled with pieces as in the following figure, then n is even. Show that the number of ways to tile a $5 \times 2k$ board with these tiles is at least $2 \cdot 3^{k-1}$.

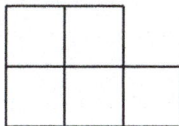

Graph Theory

4

4.1 Basic Concepts

Graph theory is one of the most useful tools when solving combinatorics problems. It helps us reduce problems and handle concepts in a much simpler way. The main tricks and ideas were presented in the previous chapters. Even though our main purpose in this book is to solve olympiad problems, we need to emphasize that graph theory is a very large and beautiful field by itself. Graph theory is said to have started in 1736, with Euler. However, the first book on graph theory was published by König in the twentieth century (1936). Even though graph theory had an accelerated growth in that century, it is still possible find open problems[1] that do not required a lot of theory to understand (and, in some cases, solve). This does not mean in any way that such problems are easy.

A **graph** G is a pair (V, E), where V is a non-empty set and E is a multiset[2] of unordered pairs of elements of V. The elements of V are called the **vertices** and the elements of E are called the **edges** of G. The reason why E is a multiset is to allow G to have an edge more than once; however, we will not deal with this kind of graphs. A **subgraph** G' of $G = (V, E)$ is a pair (V', E'), where V' is a subset of V, E' is a subset of E and (V', E') is also a graph. Given a graph G, the set of vertices is usually denoted by $v(G)$ and the (multi) set of edges is denoted by $e(G)$.[3]

Each graph has a geometric representation. One places one point in the plane for each vertex, and a pair of points are joined with a line segment whenever the

---

[1] An open problem is one that has not been solved yet.

[2] A multiset is a set than can have the same element more than once, for example $\{1, 1, 2, 3\}$ is a multiset.

[3] If the set of vertices and the set of edges are empty, this still fits the definition of a graph, commonly referred to as the **empty graph**. It is a technical detail and can normally be ignored, but it is a good thing to keep in the back of our minds when strange inductions or proofs are needed.

P. Soberón, *Problem-Solving Methods in Combinatorics*,
DOI 10.1007/978-3-0348-0597-1_4, © Springer Basel 2013

corresponding vertices form an edge. Sometimes in the geometric representation edges have a common point, this does not mean that there is a new vertex there.[4]

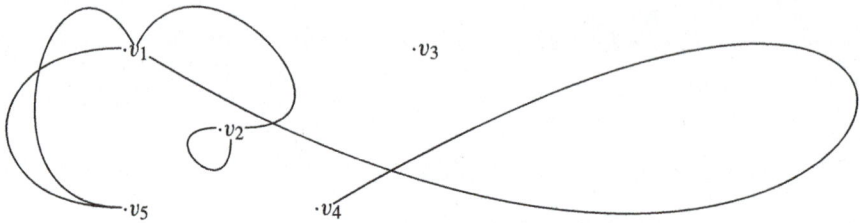

The figure shows a graph with vertices v_1, v_2, v_3, v_4, v_5 and edges $\{v_1, v_2\}$, $\{v_1, v_4\}, \{v_1, v_5\}, \{v_1, v_5\}, \{v_2, v_2\}$.

Given a graph G, we say G is **simple** if there are no edges of only one vertex and there are no multiple edges. From now on, all graphs are simple unless specified otherwise. The term **multigraph** usually refers to graphs that are not simple.

Given a graph G and its vertex v_0, we say that an edge A is **incident** in v_0 if v_0 is one of its vertices. We also say that two vertices v_0 and v_1 are **adjacent** if they form an edge. We say that v_0 has degree k if it is adjacent to exactly k vertices, and then we write $d(v_0) = k$.

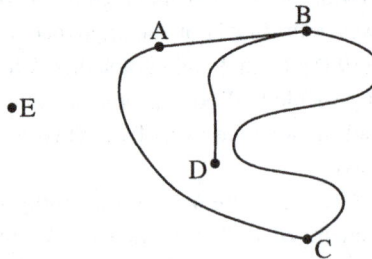

In the graph above $d(A) = 2, d(B) = 3, d(C) = 2, d(D) = 1, d(E) = 0$.

Example 4.1.1 Let G be a graph with vertex set $V = \{v_1, v_2, \ldots, v_n\}$. Then $\sum_{k=1}^{n} d(v_k) = 2|E|$.

Solution Consider the pairs (v, e) such that e is an edge of G and v is a vertex of e. The vertex v_i is in $d(v_i)$ pairs, so the total number of pairs is $\sum_{k=1}^{n} d(v_k)$. However, every edge is in 2 pairs, one for each of its vertices. So the total number of pairs is $2|E|$, as claimed.   □

---

[4]In graphs with an infinite number of vertices it may happen that the number of vertices is larger than the number of points in the plane. However, in these cases the geometric representation is clearly not important.

Exercise 4.1.2 Let G be a graph with an odd number of vertices. Show that there is at least one vertex with even degree.

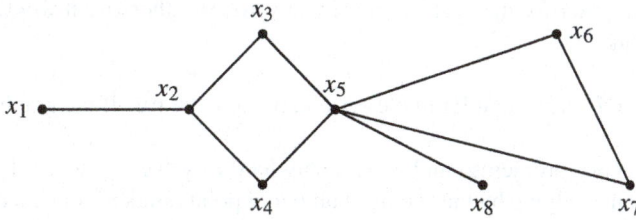

In a graph, a **walk** is a sequence of vertices $(v_0, v_1, v_2, \ldots, v_k)$ such that v_i is adjacent to v_{i+1} for all $1 \leq i \leq k - 1$. In a walk, vertices may be repeated. A **path** is a walk where no vertices are repeated. In the previous graph, $(x_1, x_2, x_3, x_5, x_6, x_7, x_5, x_4)$ is a walk, but not a path. However, $(x_1, x_2, x_3, x_5, x_4)$ is a path. We say that a walk is closed if the first and last vertices are the same. A **cycle** is a closed walk where the only two vertices that coincide are the first and the last. When we write $(v_0, v_1, \ldots, v_k = v_0)$ we mean the closed walk (or cycle) that goes through the vertices $v_0, v_1, \ldots, v_{k-1}$ and then returns to the vertex v_0. In the previous graph, $(x_4, x_2, x_3, x_5, x_7, x_6, x_5, x_4)$ is a closed walk but not a cycle. However, (x_4, x_2, x_3, x_4) is a cycle. Two cycles (walks, closed walks, paths) are said to be disjoint if they share no vertices. The length of a path is equal to the number of edges it has or, equivalently, the number of vertices it has minus one. The length of a cycle is equal to the number of edges it has or, equivalently, the number of vertices it has.

Exercise 4.1.3 Let G be a graph and u, v two of its vertices. Prove that if there is a walk that starts in u and ends in v there must be a path that starts in u and ends in v.

Exercise 4.1.4 Let G be a graph such that all its vertices have degree 2. Prove that G is a union of pairwise disjoint cycles.

Exercise 4.1.5 Consider a graph G and v_1, v_2 two of its vertices. We know that there are two walks from v_1 to v_2, one of odd length and one of even length. Show that in G there is at least one cycle of odd length.

   Note: The walks may not be disjoint.

Example 4.1.6 In a graph G every vertex has degree at least $k \geq 2$. Show that G has a cycle of length at least $k + 1$.

Solution We construct a sequence of vertices $(v_0, v_1, v_2, \ldots)$ in the following way: v_0 is any vertex and, if we have constructed $v_0, v_1, \ldots, v_{t-1}$, then we choose v_t to be adjacent to v_{t-1} and different from $v_{t-2}, v_{t-3}, \ldots, v_{t-k}$. This can be done since $d(v_{t-1}) \geq k$. Since G is a finite graph, the sequence cannot go on indefinitely without repeating vertices: there must be two vertices v_t and v_{t-l} such that $v_t = v_{t-l}$.

We can suppose that t is the first moment when this happens. Given the construction of the sequence, we have that $l \geq k + 1$. Thus $(v_{t-l}, v_{t-l+1}, \ldots, v_{t-1}, v_t = v_{t-l})$ is the cycle we were looking for. It is important that t was the first moment when this happened in order for the cycle not to repeat vertices (otherwise it would simply be a closed walk).   □

Exercise 4.1.7 Prove that Example 4.1.6 may fail to be true if the graph is infinite.

Graph theory problems can be extremely hard to solve. Up to 2011 the hardest IMO problem (judging by the average number of points students obtained) was from this category. It is the following one.

Example 4.1.8 (IMO 2007) In a mathematical competition some competitors are friends. Friendship is always mutual. Call a group of competitors a *clique* if each two of them are friends. The number of members of a clique is called its *size*. Given that, in this competition, the largest size of a clique is even, prove that the competitors can be arranged in two rooms such that the largest size of a clique contained in one room is the same as the largest size of a clique contained in the other room.

The idea behind the solution is to move students one by one from one room to another and vice versa to obtain the desired result. It should be noted that almost no technical knowledge is needed, but that is not where the difficulty of the problem lies. The solution breaks down into many cases, and obtaining a valid proof is cumbersome.

Solution We approach the problem by considering a graph that has a vertex for each student and an edge for every friendship. Denote the rooms by X and Y. Let K be a maximal clique and suppose that it has $2k$ vertices. We place the vertices of K in X and the rest in Y. In this situation the size of the largest clique in X is $2k$ and the size of the one in Y is at most $2k$. If they have the same size we have the desired partition and we stop. If not, X has a larger clique. We start to move the vertices from X to Y one by one. Note that at every step the size of the largest clique in X is reduced by 1 and the size of the largest clique in Y increases by at most one. We do this as long as the X has a larger size of maximal clique than Y. When this is no longer the case, if we do not have the situation we want, then Y has a larger size of maximal clique than X. Given the way these numbers could change, they differ by 1. That is, Y has a maximal clique of size $r + 1$, X has a maximal clique of size r.

Note that Y can have at most k vertices of K. Indeed, otherwise X would have at most $k - 1$ vertices. By analyzing Y we get $r + 1 \geq k + 1$ and by analyzing X we get $r \leq k - 1$, a contradiction. Thus X has at least k vertices of K and $r \geq k$. Suppose there is a vertex $y \in Y \cap K$ that is not in a maximal clique of Y. Then if we move Y back to X, it increases it size of the maximal clique in that room (since it is part of K). However, since there was a maximal clique of Y that did not contain Y, removing Y does not decrease the maximal size of a clique in Y. This would mean that both rooms have maximal clique size of $r + 1$, and we are done.

Thus, for every maximal clique M of Y, $Y \cap K \subset M$. However, we know that $Y \cap K$ has at most k vertices of K. Thus every maximal clique of Y must have vertices that are not in $Y \cap K$. As long as the maximal clique of Y is of size $r+1$, we do the following. Take a maximal clique M in Y and move a vertex of $M \backslash (Y \cap K)$ to X. At the end of this process the maximal size of a clique in Y is r. Let N be a maximal clique of X. Note that the vertices of N could be either vertices of K or vertices that we moved back to X in this process. Either way they are adjacent to all the vertices of $Y \cap K$. This means that $N \cup (Y \cap K)$ is a clique. Since K was a maximal clique, we have that

$$2k \geq \left| N \cup (Y \cap K) \right| = |N| + |Y \cap K|$$

Note that $2k = |K| = |X \cap K| + |Y \cap K| = r + |Y \cap K|$. Thus the last inequality translates to

$$r \geq |N|$$

This means that both X and Y have maximal clique size r, as we wanted. $\square$

The term **clique** is used in general to refer to a subgraph in which all the possible edges are present. Graphs in which all the possible edges are present are called **complete graphs**. Usually the complete graph with t vertices is denoted by K_t.

4.2   Connectedness and Trees

We say that a graph G is **connected** if given any two vertices v_0 and v_1 of G there is a walk that starts in v_0 and ends in v_1. In other words, we can get from any vertex in G to any other vertex in G by moving only along edges in the graph. By Exercise 4.1.3, one can equivalently define connectedness by changing the word "walk" by "path". We consider the graph with only one vertex and the empty graph to be connected.

Exercise 4.2.1 Let G be a connected graph and $(v_0, v_1, \ldots, v_k = v_0)$ a cycle in G. Prove that, if we remove from G the edge $\{v_0, v_1\}$, what is left is still a connected graph.

If G is a connected graph without cycles, we say that G is a **tree**. Trees are the simplest type of connected graphs, so studying them allows us to have a better understanding of the properties of connected graphs in general.

Proposition 4.2.2 *Every connected graph G has a tree T that uses all its vertices.*

Proof Let G be a connected graph. If G has no cycles, then G is a tree and we are done. If G has a cycle $(v_0, v_1, \ldots, v_k = v_0)$, we can remove the edge (v_0, v_{k-1}) from G. By Exercise 4.2.1, the subgraph we obtain is connected. We are going to prove this last statement to show how this is usually done. Let h_1, h_2 be two vertices of G. Since G is connected, there is a walk that goes from h_1 to h_2. If that walk does not use the edge $\{v_0, v_{k-1}\}$, then it is a walk in the subgraph. If it does use that edge, we can replace it by $(v_0, v_1, \ldots, v_{k-1})$ (or backwards) every time it was used. Thus we obtain a walk in the subgraph. With this we are no longer using the edge $\{v_0, v_{k-1}\}$, so the subgraph is connected. If we repeat this argument until no cycles remain, we obtain the tree we were looking for.   □

The trees that satisfy this condition are called **spanning trees** of G. A connected graph can have more than one spanning tree.

Example 4.2.3 Prove that every tree with n vertices has exactly $n - 1$ edges.

Solution We prove this by induction on n. If $n = 1$, the assertion is clear. Suppose it is true for n and we want to prove it for $n + 1$. We want to show that there is a vertex of degree 1. Note that since $n + 1 > 1$, in order for the graph to be connected every vertex must have degree at least 1. One way to show that there is a vertex of degree at most 1 is to consider the longest possible path $(v_1, v_2, \ldots, v_k)$ in the graph. If v_1 only makes an edge with v_2, we are done. If it makes an edge with any other v_i, we have a cycle, so the graph would not be a tree. If it makes an edge with any other vertex, adding this new vertex to the path contradicts its maximality. Now that we have a vertex v_1 of degree 1, we can remove it from the graph along with its edge. The new graph G' has n vertices. Note that there are no cycles in G', since G' is a subgraph of G. To see that G' is connected, consider two of its vertices. Since G is connected, there is a path in G that joins them. This path could not have used v_1, since every vertex in a path has degree greater than or equal to 2. Thus G' is a tree with n vertices, which means it has exactly $n - 1$ edges. Thus G has n edges.   □

Exercise 4.2.4 Prove that every connected graph with n vertices has at least $n - 1$ edges. There are exactly $n - 1$ edges if and only if the graph is a tree.

Exercise 4.2.5 Prove that if G is a graph with n vertices, $n - 1$ edges and no cycles, then G is connected.

Exercise 4.2.6 Show that every tree (with at least two vertices) has at least two vertices of degree 1.

Example 4.2.7 Let G be a connected graph where every vertex has degree greater than or equal to 2. Show that there are two adjacent vertices v_1, v_2 such that, if we remove them, the remaining graph is connected.

Solution Let T be a spanning tree of G. Given two vertices v_1, v_2 in G, there is a unique path that goes from v_1 to v_2 in T (if there were two, there would be a cycle). Let $P = (v_1, v_2, \ldots, v_k)$ be the longest path in T. Let $u_1, u_2, \ldots, u_l$ the vertices adjacent to v_2 in T different from v_1 and v_3. Note that the degree in T of

$v_1, u_1, u_2, \ldots, u_l$ is 1. If that is not true, we can construct a path in T longer than P, contradicting its maximality. Thus, if we remove any vertices of $v_1, u_1, u_2, \ldots, u_l$ the connectedness of T (and thus of G) is not broken. If any two of those vertices are adjacent in G, we are done. If not, remember that every vertex in G has degree at least 2, so every vertex of $u_1, u_2, \ldots, u_l$ must be adjacent (in G) to a vertex different from $v_1, v_2, u_1, u_2, \ldots, u_l$. Thus, by removing v_1, v_2 we are not breaking the connectedness of G. $\qquad\square$

In the previous example, we know that there is a cycle, by Exercise 4.2.6. However, since we want to remove adjacent vertices (not only the edge between them), we need spanning trees to attack this problem.

Exercise 4.2.8 Prove that in any connected graph G with at least three vertices there are two vertices such that if we remove any of them or both the graph remains connected.

Let G be a graph and v_0 a vertex in G. Let N_{v_0} the set of all vertices in G for which there is a walk joining them with v_0. Clearly there is walk joining any two vertices in N_{v_0}, so it is connected. N_{v_0} is called the **connected component** of G at v_0.

Exercise 4.2.9 Let v_0 and v_1 be vertices of G. Show that $v_0 \in N_{v_1}$ if and only if $v_1 \in N_{v_0}$.

Exercise 4.2.10 Let v_0 and v_1 be vertices of G. Show that if v_0 is not in N_{v_1} then N_{v_0} and N_{v_1} have no vertices in common and there are no edges between a vertex of N_{v_0} and a vertex of N_{v_1}.

Exercises 4.2.9 and 4.2.10 show that every graph can be split into connected components. Sometimes it is easier to work with the connected components of a graph than with the graph itself.

Exercise 4.2.11 Let G be a graph with no cycles, n vertices and t connected components. Show that G has exactly $n - t$ edges. What happens if $t = 1$?

Graphs with no cycles are usually referred to as **forests**. This is simply because their connected components are trees.

Given a connected graph, we can find a path between any two of its vertices. This allows us to think that the vertices in a graph have a certain distance between them.

Definition 4.2.12 Given a connected graph G, we define the distance between two different vertices u, v of G as the minimum length of a path that starts in u and ends in v. We denote this distance by $d(u, v)$. We define $d(u, u) = 0$.

Exercise 4.2.13 Show that if G is a connected graph and u, v are vertices in G, then $d(u, v) = d(v, u)$.

Exercise 4.2.14 (Triangle inequality) Show that if G is a connected graph and u, v, w are vertices in G, then $d(u, v) \leq d(u, w) + d(w, v)$.

Exercise 4.2.15 Show that if G is a connected graph and u, v, w are vertices in G, then $d(u, v) \geq |d(u, w) - d(w, v)|$.

A very good example of how these tools can be used is the second solution presented for Problem 1.14. Here we show it again.

Problem 1.14 (IMO 2006) We say that a diagonal of a regular polygon P of 2006 sides is a good segment if its extremities divide the boundary of P into two parts, each with an odd number of sides. The sides of P are also considered good segments.

Suppose that P is divided into triangles using 2003 diagonals such that no two of them meet in the interior of P. Find the maximum number of isosceles triangles with two good segments as sides that can be in this triangulation.

Solution Consider a graph G with one vertex for each of the 2004 triangles and an edge between two vertices if their triangles share a non-good side. Note that each vertex has degree 1 or 3. Good triangles have degree 1 (although vertices of degree 1 may not be isosceles!). Note that this graph cannot have a cycle, since removing any edge e makes the graph disconnected (the diagonal represented by e splits all triangles into those that are on one side and all that are on the other, and there are no edges between these two groups).

Let $N_1, N_2, \ldots, N_k$ be the connected components of G (each of them is a tree). If N_i has n_i vertices, suppose x are of degree 1 and y are of degree 3. Counting the number of vertices and the number of edges gives

$$x + y = n_i,$$
$$x + 3y = 2(n_i - 1).$$

Thus $x = \frac{n_i+2}{2}$. Hence, the total number of vertices of degree 1 is $\frac{n_1+n_2+\cdots+n_k}{2} + k = 1002 + k$. Note that the vertices of N_i represent triangles that form a polygon, and that the union of the polygons formed by all the connected components is the original 2006-gon. Consider a new graph H with one vertex for each connected component and one edge between them if their corresponding polygons share a side. Note that H is connected. Thus it has at least $k - 1$ edges. If $\triangle_1$ and $\triangle_2$ are triangles in different connected components that share a side, this side must be a good segment (as there is no edge between them!). This implies that $\triangle_1$ and $\triangle_2$ are represented in G by vertices of degree 1. Moreover, it is easy to see that at most one of them is a good triangle. Thus at least $k - 1$ vertices of degree 1 are not good triangles. This implies that there are at most 1003 good triangles. A triangulation with 1003 good triangles is obtained by considering 1003 disjoint diagonals that leave exactly 1 vertex on one side and then 1000 other diagonals to complete the triangulation.   □

Two other solutions of this problems are presented in the book in Chap. 9.

4.3 Bipartite Graphs

In a graph G, if V' is a set of vertices, we say that V' is **independent** if there are no edges that join two vertices of V'. We say that G is **bipartite** if we can split its vertices into two independent sets X and Y. X and Y are called the "components" of G (they should not be confused with the connected components). It might be possible to split a bipartite graph into components in more than one way.

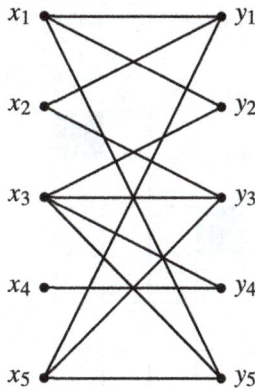

The above graph is bipartite with components $X = \{x_1, x_2, x_3, x_4, x_5\}$ and $Y = \{y_1, y_2, y_3, y_4, y_5\}$.

Exercise 4.3.1 Show that a graph G is bipartite if and only if each of its connected components is bipartite.

Exercise 4.3.2 Show that a bipartite graph G with n vertices has at most $\frac{n^2}{4}$ edges.

Theorem 4.3.3 (König 1936) *A graph G is bipartite if and only if all its cycles have even length.*

Proof We first show that, if G has a cycle of odd length, then it is not bipartite. Suppose that G has a cycle of length $2n + 1$. If G were bipartite, its vertices could be split into two independent sets X and Y. By the pigeonhole principle, at least $n + 1$ vertices of the cycle would be in X or in Y. Since among any $n + 1$ vertices of the cycle there are two consecutive ones, it is impossible for both X and Y to be independent.

Now suppose that G has no cycles of odd length. We show how to split its vertices into two independent sets. First, using Exercise 4.3.1 we can suppose that G is connected. Let X and Y be two empty sets. Choose a vertex v_0 and place it in X. Given any other vertex v_1, take a walk that begins in v_0 and ends in v_1. If the walk has even length we place v_1 in X and if it has odd length we place v_1 in Y. By Exercise 4.1.5 we know that no vertex was placed in both X and Y. If two vertices p_1 and p_2 in Y were adjacent, we can consider a walk from v_0 to p_1 and add (p_1, p_2), which gives us a walk of odd length that starts in v_0 and ends in p_2, which is impossible. In the same way we can see that X is independent. Thus G is bipartite.   □

Bipartite graphs are common when dealing with problems that involve boards. If we have an $m \times n$ board we can assign to it a bipartite graph G with $m + n$ vertices and components X, Y of m and n vertices, respectively. The vertices in X represent the columns of the board and the vertices of Y represent the rows. An edge represents the square that is in the intersection of the column and row represented by its vertices.

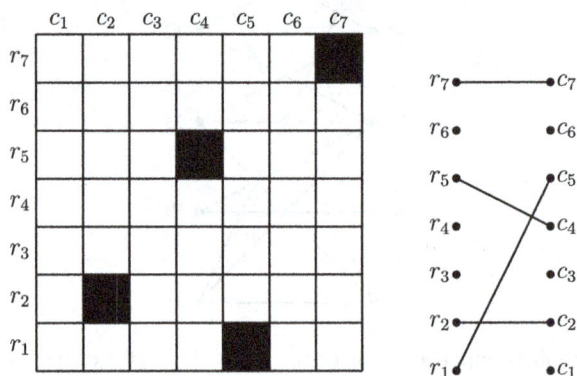

In the figure the edges represent the black squares.

Example 4.3.4 (Estonia 2007) Consider a 10×10 board. In each move, we place a token in the four squares made by the intersection of a pair of rows and a pair of columns. A move can be made if at least one of the four squares has no tokens. What is the maximum number of moves that can made if there are no tokens in the beginning?

Solution Let C_1 be the set of squares (i, j) such that $1 \le i \le 9$ and $1 \le j \le 9$. Given a square (i, j) in C_1, the move that places a token in $(10, j)$, $(10, 10)$, (i, j), $(i, 10)$ only uses that square in C_1. Thus, using only this type of moves, we can make 81 valid moves.

Now we show that 81 is the biggest possible number of moves we can make. For this, every time we make a move, we choose one of the squares that received its first token and paint it red and paint black all the others (the ones that were already painted are left untouched). In the end we have as many red squares as moves used.

Let G be a bipartite graph with 20 vertices where 10 of them form an independent set and represent the columns and the 10 others are also independent and represent the rows. We place every possible edge. This means that a coloring of the board is a coloring of the edges of G.

Let $G_T(k)$ be the subgraph formed by all the vertices and all the colored edges after making k moves. In the same way, let $G_N(k)$ be the graph formed by all the vertices and all the black edges after k moves. We prove by induction on k that the connected components of $G_T(k)$ and $G_N(k)$ are the same.

If $k = 1$ we have colored the edges (i_1, j_1), (i_1, j_2), (i_2, j_1), (i_2, j_2), of which exactly one is red. Let (i_1, j_1) be the red edge. First note that (i_1, j_2, i_2, j_1) is a walk in $G_N(1)$ that joins all the selected vertices, so the connected components of $G_N(1)$ and $G_T(1)$ are the same.

Suppose the connected components are the same in $G_T(k)$ and $G_N(k)$. Consider a new move that colors the edges (i_1, j_1), (i_1, j_2), (i_2, j_1), (i_2, j_2) (respecting what was already colored) so that (i_1, j_1) is red. Note that we may not be able to use the walk (i_1, j_2, i_2, j_1) in $G_N(k + 1)$ since some of its edges may be red. However, if any of its edges is red, then the two vertices are in the same component in $G_T(k)$. This means (by induction) that there must be a walk joining those vertices in $G_N(k)$. If we replace the red edges by these walks, we obtain a walk in $G_N(k + 1)$ that joins $\{i_1, j_2, i_2, j_1\}$. Thus, in $G_N(k + 1)$ and $G_T(k + 1)$ all we are doing is "gluing" the connected components of $G_T(k)$ these vertices were in. Thus the connected components are the same after $k + 1$ moves.

If k_0 is the last move, $G_T(k_0) = G$. This means that $G_N(k_0)$ is connected. Thus, by Exercise 4.2.4, $G_N(k_0)$ has at least 19 edges. Thus there are at most $100 - 19 = 81$ red edges, which is equal to the number of moves. $\square$

The last part of this solution (the one about $G_N(k)$ and $G_T(k)$) is very technical. However, this is a very good example of how a problem that seems to be completely unrelated to graph theory can be solved using the tools we have been developing.

4.4 Matchings

Given a graph $G = (V, E)$, we say that G' is a **matching** in G if G' is a subgraph of G such that all its vertices have degree 1. If G' is a matching, it has an even number of vertices and they are organized as disjoint adjacent pairs of vertices. Every matching G' is a bipartite graph and its components have the same number of vertices. We say that G' is a **perfect matching** of G if G' uses all the vertices of G.

Example 4.4.1 Let G be a graph with $2n$ vertices such that all the vertices have degree greater than or equal to n. Show that G has a perfect matching.

Solution We use induction on the number of pairs of vertices we have in our matching. For the base of induction, take any pair of adjacent vertices. Suppose we have a matching with r pairs, $r < n$. We will prove that we can find a matching with $r + 1$ pairs. Let A be the set of the $2r$ vertices in the matching and B the rest of the vertices of G. If there is an edge between two vertices of B, we add this pair to the matching. If not, all the edges from the vertices of B go to vertices of A. Let a and b be two vertices in B. There are $2n$ edges incident in them. In A there are r pairs of vertices. By the pigeonhole principle, at least $\lceil \frac{2n}{r} \rceil \geq 3$ edges must go to the same pair. If this pair is (u, v), the 3 edges cannot go to the same vertex. Thus one must be adjacent to a and the other to b.

If we remove the pair (u, v), we can use the vertices $\{a, b, u, v\}$ to add two adjacent pairs and increase by 1 the number of pairs in our matching. $\square$

Given a subset V' of V, the **neighborhood** of V' is defined to be the set of vertices adjacent to at least one vertex of V' and is denoted as $\Gamma(V')$. Note that if G is a bipartite graph with components X and Y, if $V' \subset X$, then $\Gamma(V') \subset Y$.

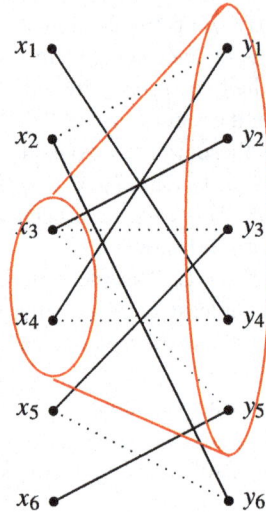

In the figure a bipartite graph with components $X = \{x_1, x_2, \ldots, x_5\}$ and $Y = \{y_1, y_2, \ldots, y_5\}$ is shown. Also, if $S = \{x_3, x_4\}$ we have that $\Gamma(S) = \{y_1, y_2, y_3, y_4, y_5\}$. The non-dotted edges of the figure form a perfect matching.

Being able to find matchings is important. The following theorem characterizes the bipartite graphs that contain a perfect matching.

Theorem 4.4.2 (Hall 1935) *Let G be a graph with components X and Y such that $|X| = |Y|$. Then G has a perfect matching if and only if $|S| \leq |\Gamma(S)|$ for all $S \subseteq X$.*

Proof First note that if G has a perfect matching, the condition is clearly satisfied. We use strong induction on $n = |X|$ to show that the theorem is true. If $n = 1$, the result is clear. Suppose that for all $k < n$ the theorem is true, and let G be a graph that satisfies the hypotheses and such that $|X| = n$. There are two cases: If there is a non-empty subset $S \subsetneq X$ such that $|S| = |\Gamma(S)|$ and if there is not.

In the first case let $T = \Gamma(S)$, $S' = X - S$ and $T' = Y - T$. Note that the graph induced by S and T satisfies the hypotheses and its components have less than n vertices, so there must be a perfect matching in that graph. Let $A \subseteq S'$. Note that $|\Gamma(S \cup A)| \geq |S \cup A| = |S| + |A| = |T| + |A|$. Thus there must be at least A vertices not in T that are adjacent to $S \cup A$. Since $T = \Gamma(S)$, these vertices are adjacent to vertices of A. Thus the graph induced by S' and T' also satisfies the hypotheses of the theorem and its components have less than n vertices. Thus it also contains a perfect matching. This implies that G contains a perfect matching.

In the second case $|\Gamma(S)| \geq |S| + 1$ for all non-empty $S \subsetneq X$. Let x be any vertex in X and y a vertex adjacent to x. Let $T = Y - \{y\}$ and $S = X - \{x\}$. Note that if $S' \subset S$ then $\Gamma(S') \geq |S'| + 1$, so there are at least $|S'|$ vertices different from y that are adjacent to S'. With this the graph generated by S and T satisfies the hypotheses of the theorem. Since its components have less than n vertices it must have a perfect matching. If we add the edge $\{x, y\}$ to this matching, we have a perfect matching in G. $\qquad\square$

The previous theorem is also know as the Marriage Theorem. This is because a common way to state it is with groups of n men and n women and giving conditions to split them into n happy marriages (if edges mean liking each other). If we remove the condition $|X| = |Y|$, we obtain a matching that uses every vertex of X instead of a perfect matching.

Example 4.4.3 (Kazakhstan 2003) We are given two square sheets of paper of area 2003. Each sheet is divided into 2003 polygons of area 1 (the divisions may be different). One sheet is placed on top of the other. Show that we can place 2003 pins such that each of the 4006 polygons is pierced.

Solution Let G be a graph G with 4006 vertices, each representing a polygon. We place an edge between two vertices if their polygons are in different sheets of paper and when one sheet is placed on top of the other these polygons have at least one point in common. Note that if S is a set of k polygons in the first sheet, they cover an area equal to k. Thus, to cover them we need at least k polygons of the second sheet of paper. This implies the condition of the Marriage Theorem in G. Thus, we have a perfect matching. Using this matching, we immediately deduce where to place the 2003 pins. $\qquad\square$

4.5   Problems

Problem 4.1 (Balkan Mathematical Olympiad 2002) Suppose G is a graph such that all its vertices have degree greater than or equal to 3. Show that G has at least one even cycle.

Problem 4.2 Let G be a graph with n vertices, such that each vertex has degree greater than or equal to $\frac{n}{2}$. Show that G is connected.

Problem 4.3 (Mantel's theorem)   Show that if G is a graph with n vertices and more than $\frac{n^2}{4}$ edges, then G has at least one triangle (a cycle of length 3).

Problem 4.4 (USAMO 1989) 20 tennis players play 14 matches, so that each one of them plays at least once. Show that there are 6 games in which 12 different players play.

Problem 4.5 (Peru 2009) At a round table there are $2n$ Peruvians, $2n$ Bolivians and $2n$ Ecuadorians. The people whose neighbors are of the same nationality are asked to stand up. What is the largest possible number of people that can stand up?

   Note: For example, for a Peruvian to stand up his two neighbors have to be of the same nationality, but they are not necessarily Peruvians.

Problem 4.6 (OMM 2009) In a party with n persons we know that out of every 4 there are 3 such that they every two of them know each other or there are 3 such that no two of them know each other. Show that we can split the persons of the party into two groups, where in one group every pair of persons know each other and in the other one no pair of persons know each other.

   Note: Knowing a person is a symmetric relation.

Problem 4.7 (Germany 2004) Given a graph with black and white vertices, a movement consists in changing the color of a vertex and every other vertex adjacent to it. If we start with a graph G with only white vertices, is it possible to turn it all black with such movements?

Problem 4.8 (Russia 2011) In a group of people some pairs are friends. A group is called k-indivisible if, for each decomposition of this group into k subgroups, at least one subgroup contains a pair of friends. Suppose that A is a finite 3-indivisible group of people having no subgroup of 4 persons such that every two of them are friends. Prove that it is possible to decompose A into two subgroups B and C such that B is 2-indivisible and C is 1-indivisible.

Problem 4.9 (IMO shortlist 2001) In a party with n persons we call k-clique a group of k persons where every two of them know each other. Suppose one knows that any two 3-cliques have a person in common and there are no 5-cliques. Show that there are 2 persons such that if they leave the party there are no 3-cliques left.

Problem 4.10 (Bulgaria 2004) There is a group of n tourists such that among any three of them there are two that do not know each other. Suppose one knows that, regardless of how we load them into two buses, there are two tourists that are in the same bus and know each other. Show that there is a tourist that knows at most $\frac{2n}{5}$ other tourists.

Problem 4.11 (Italy 2007) In a tournament there are $2n+1$ teams. Every team plays exactly once against each of the other teams. Every match ends with the victory of one of the two teams. We say that a set $\{A, B, C\}$ of 3 teams is cyclic if A beat B, B beat C and C beat A.

•   Determine the minimum number of cyclic sets in terms of n.
•   Determine the maximum number of cyclic sets in terms of n.

Problem 4.12 Given sets $S_1, S_2, \ldots, S_m$, we say they have a **system of representatives** if we can find a collection $s_1, s_2, \ldots, s_m$ of different elements such that s_i is an element of S_i for each i. Prove that $S_1, S_2, \ldots, S_m$ has a system of representatives if and only if the union of any k of them has at least k elements.

Problem 4.13 Given a bipartite graph G, show that if all the vertices have the same positive degree, then the graph has a perfect matching.

Problem 4.14 We are given an $n \times k$ board with $k < n$ such that in every square there is a number from 1 to n. We know that in every column and every row no number repeats itself. Show that we can extend this to an $n \times n$ board with a number from 1 to n in each square, such that in every row and every column no number repeats itself.

Problem 4.15 (IMO 1990) There are $2n - 1$ points in a circumference. k of them are going to be black. We say that a coloring is good if there are two black vertices such that the interior of one of the two arcs they define has exactly n vertices. Find the minimum k such that every coloring of k vertices is good.

Problem 4.16 (APMO 1989) Show that in a graph with n vertices and k edges the number of triangles is at least

$$\frac{k(4k - n^2)}{3n}.$$

Problem 4.17 (OIM shortlist 2009) 110 teams play a bridge tournament. To do this, they play 6 rounds. In each round they split into 55 pairs and each pair plays a game. We know that no pair played more than once. Show that we can find 19 teams such that no two of them played each other.

Problem 4.18 (Bulgaria 2009) In a country some cities are connected by roads. We know that, using these roads, we can get from any city to any other city, although it may not be directly. Denote by t the smallest possible integer for which there is a city from which it is possible to get to any other city using at most t roads. Show that there are cities $A_1, A_2, \ldots, A_{2t-1}$ such that, for any $1 \le i < j \le 2t - 1$, there is a road between A_i and A_j if and only if $i + 1 = j$.

Functions

5

5.1 Functions in Combinatorics

Given two sets A and B, a **function** f from A to B is a correspondence rule that "sends" each element of A to some element of B. We write $f : A \longrightarrow B$. For each element x of A, we denote by $f(x)$ the element of B to which x is sent. Consider also the set $f[A] = \{f(x) \mid x$ is in $A\}$, called the **image** of A under f. This set represents all the elements of B where we sent an element of A with f. It is important that every element of A is sent to exactly one element of B. Elements of B may not have been assigned to any element of A, or they may be assigned to more than one element. (See Fig. 5.1.)

It is common to find functions while working in problems in combinatorics. Knowing how to work with them can be of great help. Of all the functions from one set to another, we are mainly interested in three types: injective, surjective and bijective functions.

We say that a function $f : A \longrightarrow B$ is **injective** if, for all x, y in A with $f(x) = f(y)$, we have that $x = y$. In other words, a function is injective if it sends different elements to different elements.

We say that a function $f : A \longrightarrow B$ is **surjective** if, for all z in B, there is at least one element x in A such that $f(x) = z$. In other words, there is no element in B that was not assigned to an element of A. This is equivalent to $f[A] = B$.

We say that a function $f : A \longrightarrow B$ is **bijective** if it is both injective and surjective.

Given two functions $f : A \longrightarrow B$ and $g : B \longrightarrow C$, we define the **composition** $g \circ f : A \longrightarrow C$ as the function from A to C such that for all x in A, $(g \circ f)(x) = g(f(x))$.

Example 5.1.1 Prove that if $f : A \longrightarrow B$ and $g : B \longrightarrow C$ are injective, then $g \circ f : A \longrightarrow C$ is also injective.

Solution Let x and y be two elements of A such that $(g \circ f)(x) = (g \circ f)(y)$. Then $g(f(x)) = g(f(y))$. Since g is injective, this means that $f(x) = f(y)$. Since f is injective, this means that $x = y$. Thus, $g \circ f$ is injective. □

P. Soberón, *Problem-Solving Methods in Combinatorics*,
DOI 10.1007/978-3-0348-0597-1_5, © Springer Basel 2013

Fig. 5.1 The figure shows a
function f from A to B
represented by *arrows*

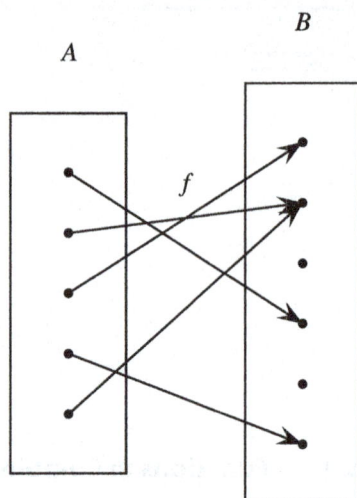

Exercise 5.1.2 *Prove that if $f : A \longrightarrow B$ and $g : B \longrightarrow C$ are surjective, then $g \circ f : A \longrightarrow C$ is also surjective.*

Exercise 5.1.3 *Given sets A and B and a function $f : A \longrightarrow B$, there is a function $g : B \longrightarrow A$ such that $(g \circ f)(x) = x$ for all x in A and $(f \circ g)(z) = z$ for all z in B if and only if f is bijective.*

Example 5.1.4 Given two finite sets A and B, prove that, if there is an injective function $f : A \longrightarrow B$, then $|A| \leq |B|$.

Solution We prove this by induction on $n = |A|$. If $n = 1$, as B must have at least one element for us to be able to define a function from A to B, the statement is true. Suppose the statement is true for every A of size n, and we want to prove it for some A of size $n + 1$. Let x be any element of A and consider the sets $A' = A \backslash \{x\}$ and $B' = B \backslash \{f(x)\}$. Note that if y is different from x, then $f(y)$ is different from $f(x)$. Thus, f sends elements in A' to elements in B'. Moreover, if we restrict f to A', it is still injective. Since $|A'| = n$, we have that $|A'| \leq |B'|$. Then, $|A| = |A'| + 1 \leq |B'| + 1 = |B|$, as we wanted. $\qquad\square$

Exercise 5.1.5 *Given two finite sets A and B, prove that, if there is a surjective function $f : A \longrightarrow B$, then $|A| \geq |B|$.*

Exercise 5.1.6 *Prove that if $f : A \longrightarrow B$ and $g : B \longrightarrow C$ satisfy that $g \circ f$ is injective, then f is also injective, but g may not be.*

Exercise 5.1.7 *Prove that if $f : A \longrightarrow B$ and $g : B \longrightarrow C$ satisfy that $g \circ f$ is surjective, then g is also surjective, but f may not be.*

Exercise 5.1.8 *Let* $f : A \longrightarrow B$ *and* $g : B \longrightarrow C$. *If* $g \circ f$ *is bijective, do any of* f *and* g *must also be bijective?*

Proposition 5.1.9 *If A and B are finite sets with the same number of elements and* $f : A \longrightarrow B$, *then the following statements are equivalent:*

- f *is injective*;
- f *is surjective*;
- f *is bijective.*

Proof

- If f is injective, then $|f[A]| = |A| = |B|$. Thus we have that $|B| = |f[A]|$ and $f[A] \subset B$. Since B is finite, this means that $B = f[A]$. That is, f is surjective.
- If f is not injective, then there are different elements x and x' in A such that $f(x) = f(x')$. Thus $|f[A]| \leq |A| - 1 < |B|$, so f cannot be surjective.

Thus f is injective if and only if f is surjective, which is equivalent to what we wanted to prove. □

Finding injective or surjective functions gives us a lot of information about the number of elements of the sets.

Exercise 5.1.10 *Given two finite sets A and B show that if there is a function* $f :$ $A \longrightarrow B$ *such that for every z in B there are exactly k elements in A whose image is z, then* $|A| = k|B|$.

Example 5.1.11 (OIM 1994) In every square of an $n \times n$ board there is a lamp. Initially all the lamps are turned off. Touching a lamp changes the state of all the lamps in its row and its column (including the lamp that was touched). Prove that we can always turn on all the lamps and find the minimum number of lamps we have to touch to do this.

First Solution First let us do the case where n is odd. If we have touched less than n lamps, we have changed at most $n - 1$ rows and $n - 1$ columns. Thus, there is at least one lamp that is still turned off. If we touch all the lamps in the first row, then all the lamps are turned on. Thus the minimum in this case is n.

If n is even, notice that by touching all the lamps once they end up all turned on. Thus the minimum is at most n^2. Also, when touching a sequence of lamps, the state of a given lamp only depends on the parity of the lamps in its row and column that were touched. With this, we can assume that no lamp was touched twice. If we can change the state of a set B of lamps, there is another set of lamps such that touching these exactly once we only change the state of B.

Let A be the family of all possible subsets of lamps. Define the function $f :$ $A \longrightarrow A$ such that $f(X)$ is the set of lamps that are on after touching exactly once every lamp of X. Since the lamps changed do not depend on the order in which the lamps of X are touched, f is well defined. Notice that after we touch all the lamps in a row and a column, the only lamp that is left on is the one in the intersection of that row and column. Given a set B of lamps, we can do this to change the state of each lamp of B and then check which lamps we actually needed to touch. This means that f is surjective. By Proposition 5.1.9, f is bijective. Thus there is only one way of turning all the lamps on (without touching a lamp more than once). Therefore, n^2 is the minimum in this case.   □

Notice that in order to obtain a bijective function for the even case we only needed to be able to change the state of a single lamp. Since in the odd case we can change the state of all the lamps in more that one way, it is impossible to change the state of a single lamp (which could be an interesting problem by itself). Another way to see that it is possible to change the state of all the lamps is to apply Problem 4.7 for the graph that represents the board.

Second Solution The case when n is odd is dealt with in the same way as in the first solution. For the even case, we know that if we touch each lamp once they will all be turned on. In the same way as in the first solution, we can suppose that each lamp has been touched at most once.

Suppose that we can change the state of all the lamps without having to touch all of them. We say that a row is odd if we touch an odd number of its squares and even otherwise. The same goes for the columns. Let L_0 be a lamp that we have not touched. In its row and its column we have touched an odd number of squares. Thus one of its row and column is odd and the other is even. Suppose the column is odd and call it c_0, the other case is analogous. Now consider any row. The square that is in this row and c_0 must change its state. Thus, there must be an odd number of lamps that were touched in them. Since c_0 is odd, there must be an even number of lamps touched in the rest of the row (without considering the intersection). Thus, there must be an even number of lamps touched outside of c_0. Adding c_0, there must be an odd number of lamps touched in total.

Now let r_0 be the row that contains L_0. Since c_0 was odd, r_0 must be even. With an argument analogous to the one in the previous paragraph, in each column there must be an odd number of lamps touched outside of r_0. Since there is an even number of columns, in total there must be an even number of lamps that were touched outside of r_0. Then the total number of lamps touched in the board must be even, which contradicts the conclusion of the previous paragraph. Thus, to change the state of all the lamps we need to touch every one of them, and the minimum is n^2.   □

Example 5.1.12 (IMO shortlist 2005) Consider an $n \times m$ board divided into nm unit squares. We say that two squares are adjacent if they share a side. By a "path" we mean any sequence of squares such that any two consecutive squares are adjacent. We say that two paths are disjoint if they do not share any square.

Each square of the board can be colored black or white. Let N be the number of colorings of the board such that there is at least one black path that goes from the left side to the right side. Let M be the number of colorings such that there are at least two disjoint black paths that go from the left side to the right side.

Show that $N^2 \geq M \cdot 2^{mn}$.

Note: In a coloring all the squares must be painted.

Solution Suppose we have a coloring of the board. Given any path a we can suppose it is not self-intersecting. If it is self-intersecting we can remove the part of the path between two repeated squares and obtain a shorter path. Thus, any path that goes from the left side to the right side divides the board into three parts (above the path, the path and below the path). The sections above the path and below the path may be empty. Let L be a path such that the upper section has maximal area. No other path can lie completely below L, and it cannot have parts below L. If it did, we could replace some part of L with the section of the other path that is below it and obtain a new path with more squares above. This also means that L is the only path with this property. We call L the lowest path of the coloring.

Now let A be the set of pairs of $n \times m$ boards colored in such a way that the first board has two disjoint black paths from left to right and the second board is colored in any way. We have that $|A| = M \cdot 2^{mn}$. Let B be the set of $n \times m$ boards colored in such a way that each of them has at least one black path that goes from left to right. Then $|B| = N^2$. Given a pair in A, let L be the lowest path of the coloring in the first board. If we exchange the upper section of that board with the corresponding squares of the second board, we obtain a pair in B. The reason is that of the two disjoint paths we had, the upper one is contained in the squares we are going to change. This defines a function $f : A \longrightarrow B$.

Notice that after applying f, L is still the lowest path of the first board, so the function may be inverted (by doing the same exchange). This means that f is injective, so $|A| \leq |B|$, which is what we wanted. $\qquad\square$

5.2   Permutations

A **permutation** σ of a set A is a bijective function $\sigma : A \longrightarrow A$. By Exercises 5.1.1, 5.1.2 and 5.1.3, a composition of permutations is also a permutation. In any set A there is at least one permutation σ_0, called the **identity in** A (even if A is empty!). It is defined as the function $\sigma_0 : A \longrightarrow A$ such that $\sigma_0(x) = x$ for all x in A. Every permutation σ has an **inverse** denoted by σ^{-1}, namely, the permutation such that

$$\sigma \circ \sigma^{-1} = \sigma^{-1} \circ \sigma = \sigma_0.$$

We denote by σ^r the permutation that results of applying σ r times. By $\sigma^{-r}(x)$ we mean $(\sigma^{-1})^r(x)$. Given two permutations σ and τ of the same set, we will write $\sigma\tau$ instead of $\sigma \circ \tau$.

In Chap. 1 we defined a permutation of a list as a way to reorder its elements. If we have an ordering of the elements of a list we can also represent it as a bijective function. To see this, think of the elements of the list as the numbers from 1 to n and define $\sigma(i)$ as the i-th element of the ordering. This is clearly a bijective function of the elements of the list. Thus, there are $n!$ bijective functions of a set of n elements into itself. A way to see this equivalence is to write the original list on top of the ordering and define the function as sending each element to the one just below,

$$\begin{pmatrix} 1 & 2 & 3 & 4 \\ 3 & 2 & 4 & 1 \end{pmatrix}.$$

Here we show how the ordering $(3, 2, 4, 1)$ of $(1, 2, 3, 4)$ corresponds to the function σ such that $\sigma(1) = 3$, $\sigma(2) = 2$, $\sigma(3) = 4$ and $\sigma(4) = 1$.

We say that x is a **fixed point** of σ if $\sigma(x) = x$. The **orbit** of x under σ is defined as the set $\mathcal{O}(x) = \{x, \sigma(x), \sigma^2(x), \sigma^3(x), \ldots\}$. That is, it is the set of all the points to which x can go if we continue to apply that same permutation. By definition, $\sigma^0(x) = x$.

Proposition 5.2.1 *Let A be a finite set and σ a permutation of A. Then for all x in A there is a positive integer h such that $\sigma^h(x) = x$.*

Proof Let n be the number of elements in A. Consider the elements x, $\sigma(x)$, $\sigma^2(x)$, $\ldots$, $\sigma^n(x)$. By the pigeonhole principle, at least two of them are equal. Let $\sigma^i(x)$ and $\sigma^j(x)$ be two equal elements of that list. If both i and j are positive, by the injectivity of σ we have that $\sigma^{j-1}(x) = \sigma^{i-1}(x)$. We can repeat this process until one of them is x and the other one is $\sigma^h(x)$ for some positive h.   □

Once we know this, for each x we are interested in the smallest positive h such that $\sigma^h(x) = x$. This number is known as the **order** of x and is denoted $o(x)$. Using this, we have that $\mathcal{O}(x) = \{x, \sigma(x), \ldots, \sigma^{h-1}(x)\}$ since, after $\sigma^{h-1}(x)$, the list repeats itself. Also, all the elements in that list are different. If A is not finite, the order of an element may not be defined. For example, if A is the set of integers and $\sigma(x) = x + 1$ for all x we have that σ is a permutation and $\sigma^t(x) = x + t$ for all t. Therefore, x will never go back to its original place in its orbit.

Exercise 5.2.2 *Show that if σ is a permutation of a finite set and $\sigma^t(x) = x$ for some t, then t is a multiple of $o(x)$.*

Exercise 5.2.3 *Show that if σ is a permutation of a finite set A and x is an element of A, then $\sigma^{-1}(x)$ is in $\mathcal{O}(x)$.*

Exercise 5.2.4 *Show that if σ is a permutation of a set A and x, y are elements of A such that $y \in \mathcal{O}(x)$ then $\mathcal{O}(y) \subset \mathcal{O}(x)$.*

Note that if we have two elements x, y of a finite set A, $y \in \mathcal{O}(x)$ if there is an r such that $\sigma^r(x) = y$. But then $x = \sigma^{-r}(y)$. If we use Exercise 5.2.3 r times, we obtain that $x \in \mathcal{O}(y)$. Therefore, x is in $\mathcal{O}(y)$ if and only if y is in $\mathcal{O}(x)$. This result allows us to prove something much stronger.

Proposition 5.2.5 *Given a permutation σ of a finite set A, any two orbits $\mathcal{O}(x)$ and $\mathcal{O}(y)$ are either disjoint or equal.*

Proof Suppose that two orbits $\mathcal{O}(x)$ and $\mathcal{O}(y)$ have a point w in common. Since w is in $\mathcal{O}(y)$ we have that y is in $\mathcal{O}(w)$. Let r be an element of $\mathcal{O}(y)$. Then we can find positive numbers k_1, k_2, k_3 such that $\sigma^{k_1}(x) = w$, $\sigma^{k_2}(w) = y$ and $\sigma^{k_3}(y) = r$. This means that $\sigma^{k_1+k_2+k_3}(x) = r$, so r is in $\mathcal{O}(x)$. Since this can be done for any element of $\mathcal{O}(y)$, we conclude that $\mathcal{O}(y)$ is a subset of $\mathcal{O}(x)$. Proving the other inclusion is analogous. □

It follows that if x is in $\mathcal{O}(y)$ then $\mathcal{O}(x) = \mathcal{O}(y)$. Thus, the set A can be partitioned into orbits.

We say that an orbit is trivial if it has only one element. In other words, it is a point x such that $\sigma(x) = x$, i.e., a fixed point of σ.

We say that a permutation is a **cycle** if it has exactly one non-trivial orbit. Cycles are the easiest permutations to analyze. They are called cycles because if we place one point in the plane for each point in A and an arrow from x to y if $\sigma(x) = y$, the picture looks just like a cycle.

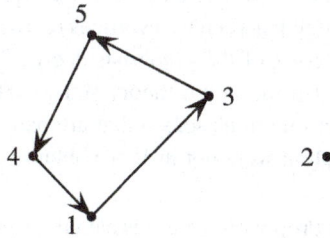

In this figure we show the graphic representation of the permutation σ of $\{1,2,3,4,5\}$ such that $\sigma(1) = 3$, $\sigma(3) = 5$, $\sigma(5) = 4$, $\sigma(4) = 1$, $\sigma(2) = 2$.

Proposition 5.2.6 *Let σ be a cycle of $\{1,2,3,\ldots,n\}$. Then there are elements $\gamma_1, \gamma_2, \ldots, \gamma_k$ of $\{1,2,\ldots,n\}$ such that*

- $\sigma(\gamma_i) = \gamma_{i+1}$, *if $1 \le i \le k-1$;*
- $\sigma(\gamma_k) = \gamma_1$;
- $\gamma_1 < \gamma_t$ *for all $t \ne 1$;*
- *If $\gamma \ne \gamma_t$ for all t, then $\sigma(\gamma) = \gamma$.*

Proof Let $\mathcal{O}(x)$ be the non-trivial orbit of the cycle and let y be the smallest element of this orbit. Then we have that $\mathcal{O}(x) = \mathcal{O}(y)$. If $h = o(y)$, then $\mathcal{O}(y) = \{y, \sigma(y), \sigma^2(y), \ldots, \sigma^{h-1}(y)\}$. Let $k = h$ and for each $1 \le i \le k$, let $\gamma_i = \sigma^{i-1}(y)$. These numbers satisfy all the required conditions. □

From now on we use the notation $\sigma = (\gamma_1, \gamma_2, \ldots, \gamma_k)$ if the γ_i are as in Proposition 5.2.6. It is clear from the proof that this way of writing σ is unique. We say that two cycles $\sigma = (\gamma_1, \gamma_2, \ldots, \gamma_k)$ and $\tau = (\alpha_1, \alpha_2, \ldots, \alpha_l)$ are **disjoint** if $\gamma_i \ne \alpha_j$ for all i, j. We also say that σ is a **transposition** if it is a cycle of the type (γ_1, γ_2). In the previous figure, we have that $\sigma = (1,3,5,4)$.

Proposition 5.2.7 *If σ and τ are disjoint cycles of the same set, then they commute. In other words, $\sigma\tau = \tau\sigma$.*

Proof Let $\sigma = (\gamma_1, \gamma_2, \ldots, \gamma_k)$ and $\tau = (\alpha_1, \alpha_2, \ldots, \alpha_l)$. If x is a fixed point of τ and σ, then $\sigma\tau(x) = x = \tau\sigma(x)$. If $x = \alpha_i$, then $\sigma\tau(\alpha_i) = \sigma(\alpha_{i+1}) = \alpha_{i+1}$ (with $\alpha_{l+1} = \alpha_1$). Also $\tau\sigma(\alpha_i) = \tau(\alpha_i) = \alpha_{i+1}$. If $x = \gamma_i$, the proof is analogous.   □

Proposition 5.2.8 *Every permutation σ of a finite set is a finite composition of pairwise disjoint cycles.*

Proof In the proof of Proposition 5.2.6 we have seen that every non-trivial orbit can be seen as a cycle. Since different orbits are disjoint, they are not modifying the same elements. By using these cycles we obtain the original permutation.   □

Second Proof (with graphs) Let V be the set of elements x such that $\sigma^2(x) \neq x$. All other points are fixed or are part of a cycle of the type (a, b). Let G be a graph with vertex set V and an edge between two vertices representing i and j if $\sigma(i) = j$ or $\sigma(j) = i$. Then every vertex of the graph has degree 2. By Exercise 4.1.4, G is a union of disjoint cycles (in the graph theory sense). These graph cycles clearly represent cycles (in the permutations sense) that are pairwise disjoint and generate σ in V. Adding the cycles that were not in V we obtain what we wanted.   □

What we established in Proposition 5.2.8 is called the **cycle decomposition** of σ. It is very important when solving problems.

If we have a permutation σ of a set A and we want to obtain its cycle decomposition, we can do the following. First, consider the smallest element x of A and start writing $x, \sigma(x), \sigma^2(x), \ldots$ until we reach the end of that orbit. This yields a cycle of the form $(x, \sigma(x), \sigma^2(x), \ldots, \sigma^{h-1}(x))$. We can repeat this process until we have all the non-trivial orbits written. For example, if $A = \{1, 2, 3, 4, 5, 6, 7\}$ and $\sigma(1) = 3$, $\sigma(2) = 2$, $\sigma(3) = 4$, $\sigma(4) = 1$, $\sigma(5) = 7$, $\sigma(6) = 5$, $\sigma(7) = 6$, we have the function represented in the following figure.

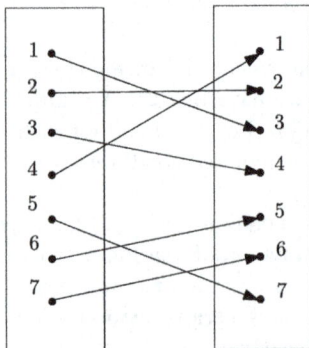

If we start with 1, we have that $\sigma(1) = 3$, $\sigma^2(1) = 4$, $\sigma^3(1) = 1$. This gives the cycle $(1, 3, 4)$. The next unused number is 2, which is a fixed point (so we ignore it). The next unused number is 5. Since $\sigma(5) = 7$, $\sigma^2(5) = 6$, $\sigma^3(5) = 5$, we get the cycle $(5, 7, 6)$. Thus $\sigma = (1, 3, 4)(5, 7, 6)$, which can be represented as the following figure.

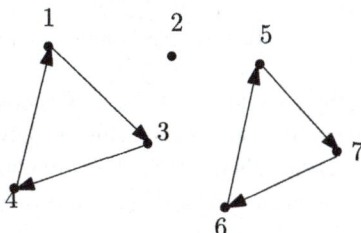

Exercise 5.2.9 *Let $A = \{1, 2, 3, 4, 5, 6\}$ and let σ and τ be two permutations of A such that $\sigma = (1, 3)(4, 5)$ and $\tau = (1, 2, 3)(5, 6)$. Compute the cycle decomposition of $\sigma\tau$. You should obtain $\sigma\tau = (1, 2)(4, 5, 6)$.*

Exercise 5.2.10 *Show that $\sigma\tau = \tau\sigma$ does not always hold if σ and τ are permutations of the same set.*

Exercise 5.2.11 *Show that if $\sigma_1, \sigma_2, \sigma_3, \ldots, \sigma_n$ are permutations of a set A, then $(\sigma_1\sigma_2 \cdots \sigma_n)^{-1} = \sigma_n^{-1}\sigma_{n-1}^{-1} \cdots \sigma_1^{-1}$.*

Example 5.2.12 Show that every cycle is a product of transpositions.

Solution Consider the cycle $\beta = (\gamma_1, \gamma_2, \gamma_3, \ldots, \gamma_k)$. Then

$$\beta = (\gamma_1, \gamma_k)(\gamma_1, \gamma_{k-1})(\gamma_1, \gamma_{k-2}) \cdots (\gamma_1, \gamma_2).$$

Let us verify that every element is sent accordingly to the cycle. To see this, note that the first transpositions applied are the ones on the right. γ_1 is first sent to γ_2, and then it does not move. For $1 < i < k$, the first time γ_i is moved is in the transposition (γ_1, γ_i), so it is sent to γ_1. Then we use the transposition (γ_1, γ_{i+1}), so it is sent to γ_{i+1}. Note that γ_k is only changed by the last transposition, and it is sent to γ_1. □

Exercise 5.2.13 *Show that every permutation is a product of transpositions.*

Exercise 5.2.14 *Show that the cycle decomposition is unique, except for the order of the factors.*

Exercise 5.2.15 *Let σ and τ be two cycles of a finite set A. Show that if their non-trivial orbits have exactly one element in common, then $\sigma\tau$ is also a cycle.*

Example 5.2.16 Find the number of permutations σ of $\{1, 2, 3, 4, 5, 6, 7\}$ such that σ^3 is the identity.

Solution Write σ using its cycle decomposition: $\sigma = \beta_1 \beta_2 \cdots \beta_k$. We have that $\sigma^3 = (\beta_1 \beta_2 \cdots \beta_k)^3$. By Proposition 5.2.7, disjoint cycles commute, so $\sigma^3 = (\beta_1)^3 (\beta_2)^3 \cdots (\beta_k)^3$. This means that β_i^3 is the identity for all i. But if $\beta_i = (\gamma_1, \gamma_2, \ldots, \gamma_k)$, then $\beta_i^3(\gamma_j) = \gamma_{j+3}$. Then, $\gamma_j = \gamma_{j+3}$ and every cycle must have length 3. Thus σ must be the identity, a cycle of length 3, or the composition of two disjoint cycles of length 3.

If it is a cycle of length 3, we have $\binom{7}{3}$ ways of choosing the elements of the cycle and 2 ways to make a cycle with them. Thus we have 70 possibilities for σ. If it is the product of two cycles, there are $\binom{7}{3}$ ways to choose the elements of one cycle and $\binom{4}{3}$ to choose the elements of the other. Since we are counting each choice twice, there are $\frac{\binom{7}{3}\binom{4}{3}}{2} = 70$ ways to choose the elements of the cycles and 2 ways to cycle each. Thus there are a total of $1 + 2(35) + 4(70) = 351$ possibilities for σ. $\square$

Example 5.2.17 (Bulgaria 2006) Let A be an $n \times n$ matrix[1] whose entries are $0, 1, -1$ and each row and column has exactly one 1 and one -1. Decide if it is possible, by successively switching the place of rows or columns, to reach to the matrix $-A$.[2]

First Solution (with permutations) Let $\sigma(j)$ be the row in which the 1 of the j-th column is, and $\tau(j)$ be the row in which the -1 of the j-th column is. By the conditions of the problem, σ and τ are permutations. Note that changing the columns i and j is the same as composing σ with the transposition (i, j) on the right and the same thing for τ. Changing the rows (i, j) is the same as composing σ with the transposition (i, j) on the left and the same thing for τ. Since every permutation is a composition of transpositions (Exercise 5.2.13), we only need to prove that there are permutations a and b such that $a\sigma b = \tau$ and $a\tau b = \sigma$.

In the first equation we have that $b = \tau a^{-1}\sigma^{-1}$. In the second one we have that $b = \sigma a^{-1}\tau^{-1}$. Thus, the permutation b can be defined if and only if $\tau a^{-1}\sigma^{-1} = \sigma a^{-1}\tau^{-1}$. The task is now reduced to finding the permutation a that satisfies this. By simple manipulation, this is equivalent to $a^{-1}\sigma^{-1}\tau a = \tau^{-1}\sigma$. If we call $\rho = \sigma^{-1}\tau$, we need a permutation a such that

$$\rho^{-1} = a^{-1}\rho a.$$

Let $\rho = \beta_1 \beta_2 \cdots \beta_r$ be its cycle decomposition. If $\beta_i = (\gamma_1, \gamma_2, \ldots, \gamma_k)$, we define $a(\gamma_j) = \gamma_{k+1-j}$ (considering the indices modulo k). Since the cycles were disjoint, a is well defined. Also, $a = a^{-1}$. Note that $(a^{-1}\rho a)(\gamma_j) = (a\rho a)(\gamma_j) =$

---

[1] A **matrix** is a rectangular array of numbers.

[2] Here $-A$ denotes to the matrix that is obtained by changing all the signs in A to the opposite ones.

$(a\rho)(\gamma_{k+1-j}) = a(\gamma_{k+2-j}) = \gamma_{k+1-(k+2-j)} = \gamma_{j-1} = \rho^{-1}(\gamma_j)$. Thus, a can always be found and for every matrix as in the problem we can find such transformation. $\square$

Second Solution (with graphs) We consider the problem in a board instead of a matrix. Let G be a graph with a vertex for each 1 and each -1 and an edge between two vertices if they are in the same column or if they are in the same row. Every vertex has degree 2, so, by Exercise 4.1.4, the graph is a union of disjoint cycles (cycles in terms of graphs). Also, cycles are alternating between 1 and -1, so they are of even length.

Let $(v_1, v_2, \ldots, v_{2k})$ be the longest cycle, with v_1 representing a 1. We can make changes of rows and columns to get the 1 from v_1 to the top left corner. Suppose that v_2 is the -1 that is in the same row (if not, the argument can be repeated with v_{2k} instead). We can exchange columns so that the -1 of v_2 is just on the right of the 1 of v_1. After that, we can make a change of columns to make the 1 of v_3 to be just below the -1 of v_2. If the cycle is not closed, the -1 of v_4 cannot be on the left side of v_3, so we can make column exchanges without modifying what we have to place it just to the right of the 1 of v_3.

We can repeat this process until we get to the -1 of v_{2k}, which is fixed in the end. We obtain a board of the following type.

1	-1			
	1	-1		
		1	-1	$\cdots$
-1			1	
		$\vdots$		$\ddots$

We repeat this process for the rest of the cycles until we have a board with this kind of stairs in the main diagonal. Since by exchanging the 1s and -1s we get the same graph, we can get the same board of stairs from both of them. Since we can get to the same board from A and $-A$, we can get from one to the other. $\square$

5.3 Counting Twice

Two sets A and B have the same number of elements if and only if there is a bijective function $\sigma : A \longrightarrow B$. So, if we want to know how many elements a given set has, it can be a good idea to find a set with the same number of elements, but that is easier to count. In other words, we want to count the set in a different way. This trick is constantly used in combinatorics problems of all levels, and some can be solved using exclusively this technique.

Example 5.3.1 Find the value of $\sum_{k=1}^{n} k\binom{n}{k}$.

First Solution Let us look for another way to see the sum we have in the example. Suppose we have a set of n persons and we want to choose some of them to form a committee with a president. If we count how many committees of this type have k persons, there are $\binom{n}{k}$ ways to choose the committee and, once we have the persons in it, there are k ways to choose the president. Thus, the sum we want is the number of possible committees (this is indirectly the bijective function we mentioned at the beginning of this section). However, we can start by choosing the president (we have n possibilities) and whether each of the $n - 1$ remaining persons is in the committee or not. This gives 2 possibilities for each other person. Thus, there are $n2^{n-1}$ possible committees, and that is the number we were looking for.   □

In this example the set "easy to count" is the set of committees and the set "difficult to count" is an arbitrary set with as many elements as the sum in the problem. This problem can be also solved with an adequate change of variables.

Second Solution Let $S = \sum_{k=0}^{n} k\binom{n}{k}$. Note that the sum can be taken starting from $k = 0$ without changing the result. Note that $S = \sum_{t=0}^{n} t\binom{n}{t} = \sum_{t=0}^{n} t\binom{n}{n-t}$. By using the change of variable $k = n - t$ in the last sum, we obtain $S = \sum_{k=0}^{n}(n - k)\binom{n}{k}$,

$$2S = \sum_{k=0}^{n}(k + [n - k])\binom{n}{k}$$

$$= n\sum_{k=0}^{n}\binom{n}{k}$$

$$= n \cdot 2^n.$$

Thus $S = n \cdot 2^{n-1}$, as in the first solution.   □

Example 5.3.2 (IMO shortlist 2004) There are 10001 students in an university. The students gather in gangs (a student may be in several gangs) and the gangs gather in societies (a gang may be in several societies). Suppose there are k societies and the following conditions are met:

- Every pair of students is in exactly one gang.
- For each student and each society, the student belongs to exactly one gang of the society.
- Each gang has an odd number of students. Also, if a gang has $2m + 1$ students (with m a non-negative integer), then it is in exactly m societies.

Find all possible values of k.

Solution Let a be a fixed student. To solve this problem, let us count the number of pairs (G, S) such that S is a society, G is a gang such that a is in G and G is in S. By the second condition, for each society there is exactly one of these pairs, so there are

k pairs. Now, let us count this number using the gangs. If a is in gang P_0, there must be another $2m$ persons in P_0 for some non-negative m. Thus, we have m pairs of the type (P_0, S). Since a is in exactly one gang with each of the other 10000 students, as we change of P_0 we obtain exactly 5000 of these pairs. This gives $k = 5000$. Also, if there is only one gang with the 10001 students and 5000 identical societies (each with the chosen gang) it is clear that all conditions are met. Thus $k = 5000$ is the only possible value.   □

In the previous example the set that was hard to count was the one of the societies and the set easy to count was the one of the pairs we considered.

Example 5.3.3 (Argentina 1997) There are n cars, numbered from 1 to n, and a row with n parking spots, numbered from 1 to n. Each car i has its favorite parking spot a_i. When it is its turn to park, it goes to his favorite spot. If it is free it parks, and if it is taken it advances until the next free spot and parks there. If it cannot find a parking spot this way, it leaves and never comes back. First car number 1 tries to park, then car number 2 and so on until car number n. Find the number of lists of favorite spots $a_1, a_2, \ldots, a_n$ such that all the cars park.
  Note: Different cars may have the same favorite spot.

Solution Let us think of a new situation. Now there are going to be $n + 1$ parking spots in a circle. The cars can also have as favorite spot the place $n + 1$, and the rules for parking are the same (the cars park in the first free parking spot starting clockwise from their favorite one). In this way, every car parks and there is always one free spot. Note that if we have a list of favorite spots $a_1, a_2, \ldots, a_n$ that leave free the spot j then the list $a_1 + 1, a_2 + 1, \ldots, a_n + 1$ leaves free the spot $j + 1$ (everything modulo $n + 1$).
  Then for each list that leaves free the spot j there is a list that leaves free the spot $j + 1$. Since the operation we did can be reversed (by subtracting 1 modulo $n + 1$), we have a bijective function between the lists that leave free the spot j and the lists that leave free the spot $j + 1$. Repeating this for each spot there is the same number of lists that leave it free. There is a total of $(n + 1)^n$ lists, so there must be exactly $(n + 1)^{n-1}$ lists that leave free the spot $n + 1$. Now notice that the good lists in the original situation and the lists in this situation that leave free the spot $n + 1$ are the same. This is because using a list of the original problem where all the cars are parked allows no car to get to spot $n + 1$ in the second situation. In a list where no car parks in spot $n + 1$ we have that $n + 1$ was not the favorite spot of any car, and they all had to park before getting there. Thus, this is also a good list in the original problem. Therefore we have a total of $(n + 1)^{n-1}$ good lists.   □

In the previous example the set that was difficult to count was the one asked for and the set easy to count was that of lists in the new situation. As we have seen, this technique is a very good way to attack problems, some of which require very little additional work. The main problem with this technique is that there is no way to make it systematic. That is, there is no recipe allowing us to always find a bijective function that works. The only way to be able to use this technique appropriately is to have done enough problems of this kind so that the bijective functions come out naturally.

5.4   The Erdős-Ko-Rado Theorem

The Erdős-Ko-Rado theorem is concerned with the size of families of intersecting sets. We include it in this chapter because there is a very simple and elegant proof of the theorem by Katona using a double counting argument.[3] Let us start with the following example.

Example 5.4.1 Let $\mathcal{F}$ be a family of subsets of $\{1, 2, \ldots, n\}$ such that every two sets in $\mathcal{F}$ are intersecting. Show that $\mathcal{F}$ has at most 2^{n-1} subsets.

Solution Consider all subsets of $\{1, 2, \ldots, n\}$. We can split them into pairs (A, B) such that A is the complement of B. Notice that in each pair we may choose at most one element for $\mathcal{F}$, so we can have at most half of all subsets. The number 2^{n-1} can be achieved if $\mathcal{F}$ is the family of subsets that contain $\{n\}$, for example. $\square$

One can ask what happens if we require that all sets if $\mathcal{F}$ have exactly k elements. We consider the case $k \leq \lfloor \frac{n}{2} \rfloor$, since otherwise the answer is immediately $\binom{n}{k}$. Notice that if we consider the family of all sets of size k that contain a fixed element, then we have $\binom{n-1}{k-1}$ sets that satisfy the condition. The Erdős-Ko-Rado theorem states that this is indeed the best possible number of sets. Namely,

Theorem 5.4.2 (Erdős, Ko, Rado 1961) *Let $\mathcal{F}$ be a family of subsets of size k of $\{1, 2, \ldots, n\}$ such that every two sets in $\mathcal{F}$ are intersecting. If $k \leq \lfloor \frac{n}{2} \rfloor$, then*

$$|\mathcal{F}| \leq \binom{n-1}{k-1}.$$

The proof of this theorem involves placing the elements in a cycle and using a double counting argument. For this we need the following result:

Proposition 5.4.3 *Let X be a set of n points in a circle and A be a family of intervals of k consecutive elements each. If every two intervals in A are intersecting and $k \leq \lfloor \frac{n}{2} \rfloor$, then A has at most k intervals.*

---

[3]The presentation of this proof is following the lines of the paper: Brouwer, A. E., Schrijver, A. **Uniform hypergraphs**, pp. 39–73, in Packing and Covering in Combinatorics, Mathematical Centre Tracts, 106, Math. Centr., Amsterdam, 1979. The interested reader may find references and many related problems therein.

Proof In what follows, consider the elements of X in clockwise order. For each interval in A we are going to paint blue its last element, and red the element that is just behind its first element, as in the picture below

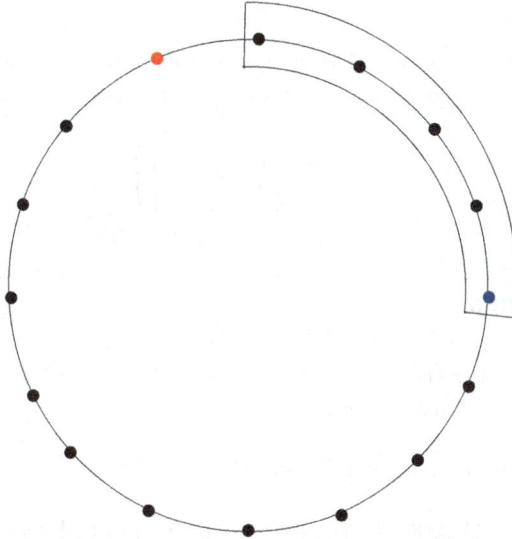

Note that the set of blue points is obtained from the set of red points by a rotation. Also, after a blue point, the following $n - 2k$ points cannot be red. That is because if any of them was red, its interval would be disjoint from the interval of the blue point we considered. Since the blue points are obtained from the red points by a rotation, this means that we can find an interval of length $n - 2k$ with no blue points. We can move this interval counter-clockwise, until the first moment when its first element is following a blue point. At this moment, it contains no blue points and, as it is following a blue point, contains no red points. Thus all blue and red points must be among the $2k$ points outside this interval. Notice that no point is painted with both colors, since the intervals they define would then be disjoint. Since there is the same number of blue and red points, there are at most k of each, as we wanted.   □

Now we are ready to prove Theorem 5.4.2.

Proof Let A be the number of pairs (I, τ) such that τ is a way of placing $\{1, 2, \ldots, n\}$ in a circle[4] and I is a set of $\mathcal{F}$ such that, in τ, I is an interval of k consecutive elements. Let B be the number of pairs (I, τ) as in A, but without re-

---

[4]We are only interested in the cyclic order, not the actual position of each point.

quiring I to be a set of $\mathcal{F}$. Note that, as we vary τ, both in A and B we are counting every set of k elements the same number of times. Call that number p. Then

$$A = |\mathcal{F}|p, \qquad B = \binom{n}{k}p.$$

However, every τ is in at most k pairs of A (by the proposition above) and is in n pairs of B. Thus $A \le \frac{k}{n}B$. Combining these two equations we obtain

$$|\mathcal{F}| \le \frac{k}{n}\binom{n}{k} = \binom{n-1}{k-1}.$$

$\square$

5.5   Problems

Problem 5.1 Let $A = \{1, 2, \ldots, n+1\}$ and $B = \{1, 2, \ldots, n\}$. Find the number of functions $f : A \longrightarrow B$ that are surjective.

Problem 5.2 Find the value of $\sum_{k=2}^{n} k(k-1)\binom{n}{k}$.

Problem 5.3 (APMO 2006) In a circus there are n clowns. Every clown may dress or paint himself using at least 5 out of 12 possible colors. We know that there are no two clowns with exactly the same colors and that no color is used by more than 20 clowns. Find the largest possible value of n.

Problem 5.4 (OIM 2003) We are given two lists of 2003 consecutive integers each and a 2×2003 board. Decide if it is possible to arrange the numbers of the first list in the first row and the numbers of the second list in the second row (without using the same number more than once) so that the sum by columns forms a new list of 2003 consecutive integers. Do the same thing but with 2003 replaced by 2004.

Problem 5.5 (Estonia 2007) An exam with k question is presented to n students. A student fails the exam if he gets less than half the answers right. We say that a question is easy if more than half of the students get it right. Decide if it is possible that

- All students fail even though all the questions were easy.
- No student fails even though no question was easy.

Problem 5.6 (Iran 2011) A school has n students and there are some extra classes they can participate in. A student may enter any number of extra classes. We know there are at least two students in each class. We also know that if two different classes have at least two students in common, then their numbers of students are different. Show that the total number of classes is not greater than $(n-1)^2$.

Problem 5.7 Show that if m, n are positive integers, then

$$\sum_{k=0}^{n}(-1)^k\binom{m+1}{k}\binom{m+n-k}{n-k}=0.$$

Problem 5.8 Show that two permutations σ and τ of a finite set A have the same cycle structure (the same number of cycles in their cycle decomposition, with equal corresponding lengths) if and only if there is a permutation α of A such that $\alpha\sigma\alpha^{-1}=\tau$.

Problem 5.9 (Switzerland 2010) In a village with at least one person there are several associations. Each person of the village is a member of at least k associations and any two different associations have at most one member in common. Show that there are at least k associations with the same number of members.

Problem 5.10 Let X be the set of permutations of $\{1, 2, \ldots, n\}$ and Y the set of functions $g : \{1, 2, \ldots, n\} \longrightarrow \{0, 1\}$. Define $T : X \longrightarrow Y$ so that each f in X is sent to the function $T_f : \{1, 2, \ldots, n\} \longrightarrow \{0, 1\}$ with the property that $T_f(a) = 1$ if and only if $f^{12}(a) = a$. Find $\sum_{f\in X}\sum_{i=1}^{n}T_f(i)$.

Problem 5.11 (IMO 1987) Let n be a positive integer and $P(k)$ be the number of permutations of $\{1, 2, \ldots, n\}$ that have exactly k fixed points. Show that

$$\sum_{k=0}^{n}kP(k)=n!.$$

Problem 5.12 (IMO shortlist 2001) Let n be a positive integer. We say that a sequence of zeros and ones is balanced if it has exactly n ones and n zeros. Two balanced sequences a and b are neighbors if we can move one of the $2n$ symbols of a to another position to obtain b. Show that there is a set S of at most $\frac{1}{n+1}\binom{2n}{n}$ balanced sequences such that every balanced sequence is in S or has a neighbor in S.

Problem 5.13 (IMO 1988) Let n be a positive integer and let A_1, A_2, A_3, $\ldots$, A_{2n+1} be subsets of a set B. Suppose that

- Every A_i has exactly $2n$ elements.
- Every intersection $A_i \cap A_j$ $(1 \leq i < j \leq 2n + 1)$ has exactly one element.
- Every element of B belongs to at least two of the sets A_i.

For which values of n can you assign to each element of B one of the numbers 0 or 1 such that every A_i has 0 assigned to exactly n of its elements?

Problem 5.14 (IMO 1998) In a competition there are m contestants and n judges, where $n \geq 3$ is an odd number. Every judge evaluates every contestant as "satisfactory" or "non-satisfactory". Suppose that every pair of judges agrees for at most k contestants. Show that

$$\frac{k}{m} \geq \frac{n-1}{2n}.$$

Problem 5.15 Consider a family $A_1, A_2, \ldots, A_k$ of subsets of $\{1, 2, \ldots, n\}$ such that for every $i \neq j$, A_i is not a subset of A_j. Show that $k \leq \binom{n}{\lfloor \frac{n}{2} \rfloor}$.

Problem 5.16 (OIM 2008) In a match of biribol two teams of 4 persons each play. A tournament of biribol is organized by n persons in which they form teams to play (the teams are not fixed). At the end of the tournament we know that every pair of persons played exactly once in opposite teams. Find all the possible values for n.

Problem 5.17 (OMCC 2004) Necklaces are made with pearls of different colors. We say that a necklace is prime if it cannot be split into chains of pearls of the same length and with identical sequence of colors. Let n and q be positive integers. Show that the number of prime necklaces with n pearls, each with one of q^n possible colors, is equal to n times the number of prime necklaces with n^2 pearls, each with one of q possible colors.

Note: Two necklaces are considered equal if they have the same number of pearls and one of them can be rotated to match the other one.

Problem 5.18 (IMO 2008) Let n and k be positive integers such that $k \geq n$ and $k - n$ is even. There are $2n$ lamps numbered $1, 2, \ldots, 2n$, and each of them can be either off or on. Initially all the lamps are turned off. Consider sequences of steps, where in each step exactly one lamp is selected and its state is changed. Let N be the number of sequences of k steps by the end of which the lamps $1, 2, \ldots, n$ are all on and the lamps $n + 1, \ldots, 2n$ are all off. Let M be the number of sequences of k steps by the end of which the lamps $1, 2, \ldots, n$ are all on and the lamps $n + 1, \ldots, 2n$ are all off without having ever been turned on. Find $\frac{N}{M}$.

Problem 5.19 (IMO 1989) Let n and k be positive integers. S is a set of n points in the plane such that:

- No 3 points of S lie in a line.
- For each point P of S there are at least k points of S with the same distance to P.

Show that $k < \frac{1}{2} + \sqrt{2n}$.

Generating Functions

6

6.1   Basic Properties

Example 6.1.1 (Great Britain 2007) In Hexagonia, there are six cities connected by railways in such a way that between every two cities there is a direct railway. On Sundays, some railways close for repairs. The train company guarantees that passengers will be able to get from any city to any other city (it may not be in a direct way) at any time. In how many ways can the company close railways so that this promise is kept?

Solution In graph theory language, we want to count the number of connected graphs that use six distinguishable given vertices. To do this counting we consider the problem with n vertices and call $f(n)$ the number we want.

Note that the total number of graphs with n vertices is $2^{\binom{n}{2}}$. Let us count these graphs in a different way.

Let v be any vertex. If the connected component in which v is has k vertices, there are $\binom{n-1}{k-1}$ ways to choose the other $k-1$ vertices of that component. With this, there are $f(k)$ ways to put the edges of that component and $2^{\binom{n-k}{2}}$ ways to put edges in the rest of the graph. We obtain

$$2^{\binom{n}{2}} = \sum_{k=1}^{n} \binom{n-1}{k-1} f(k) \cdot 2^{\binom{n-k}{2}}.$$

This is a recursive formula. It can be used to compute $f(6)$. First we have that $f(1) = f(2) = 1$. Using the formula for $n = 3$ to find $f(3)$ gives $f(3) = 4$. With the same idea, $f(4) = 28$, $f(5) = 728$ and $f(6) = 26704$. However, since the company has to close at least one railway, the number we wanted is 26703 (eliminating the complete graph with 6 vertices). □

Many times in combinatorics problems we find sequences of numbers and only recursive equations to work with them. Generating functions are a tool that allows

P. Soberón, *Problem-Solving Methods in Combinatorics*,
DOI 10.1007/978-3-0348-0597-1_6, © Springer Basel 2013

us to work efficiently with these equations. They are very useful to find **closed-form formulas,** i.e., formulas that do not depend on previous terms of the sequence.

What we do is to assign to each sequence a function in the following way:

$$(a_0, a_1, a_2, a_3, \ldots) \longleftrightarrow f(x) = a_0 + a_1 x + a_2 x^2 + a_3 x^3 + \cdots.$$

The function $f(x)$ is called the **generating function** of the sequence $(a_0, a_1, a_2, \ldots)$ and represents this sequence. a_0 is called the **independent term** of $f(x)$. We have to note that $f(x)$ is, at this point only, a way to refer to the sequence, so we are not interested in evaluating it at some point. When can generating functions be evaluated at a point will be explained in Sect. 6.5.

We can sum and multiply generating functions. Given two functions $f(x)$ and $g(x)$ associated to the sequences $(a_0, a_1, a_2, \ldots)$ and $(b_0, b_1, b_2, \ldots)$, respectively, the sum and product are defined by

$$(f+g)(x) = f(x) + g(x) = (a_0 + b_0) + (a_1 + b_1)x + (a_2 + b_2)x^2 + \cdots,$$
(6.1)

$$(fg)(x) = f(x)g(x)$$
$$= (a_0 b_0) + (a_0 b_1 + a_1 b_0)x + (a_0 b_2 + a_1 b_1 + a_2 b_0)x^2 + \cdots. \quad (6.2)$$

In the product, the coefficient of x^k is $a_k b_0 + a_{k-1} b_1 + a_{k-2} b_2 + \cdots + a_1 b_{k-1} + a_0 b_k$. With these rules, generating functions behave like infinite polynomials.

Proposition 6.1.2 *For any generating function $f(x)$ with independent term different from 0 there is a unique generating function $g(x)$ such that $f(x)g(x) = 1$.*

Proof We construct the terms of the sequence associated to $g(x)$ in an inductive way. Let $(a_0, a_1, a_2, \ldots)$ and $(b_0, b_1, b_2, \ldots)$ the sequences associated to $f(x)$ and $g(x)$, respectively. We know that $f(x)g(x) = (a_0 b_0) + (a_0 b_1 + a_1 b_0)x + (a_0 b_2 + a_1 b_1 + a_2 b_0)x^2 + \cdots$. We want $a_0 b_0 = 1$. Thus, $b_0 = \frac{1}{a_0}$. Suppose that up to some k we have constructed $b_0, b_1, \ldots, b_k$ and look at the $(k+1)$-th term of $f(x)g(x)$. Since we want the product to be 1, all the coefficients except for the independent term must be 0. However, we know that the coefficient of x^{k+1} is $a_{k+1} b_0 + a_k b_1 + a_{k-1} b_2 + \cdots + a_1 b_k + a_0 b_{k+1}$. Solving for b_{k+1}, we get

$$b_{k+1} = \frac{-(a_{k+1} b_0 + a_k b_1 + a_{k-1} b_2 + \cdots + a_1 b_k)}{a_0}.$$

So b_{k+1} is uniquely determined.   □

The function $g(x)$ is denoted by $\frac{1}{f(x)}$. This helps us a lot when working with generating functions. If the independent term of $f(x)$ is 0, then $f(x) = xh(x)$ for some generating function $h(x)$. Thus, for any generating function $g(x)$, we have that $f(x)g(x) = xh(x)g(x)$. All the terms of this product are multiples of x, so we cannot make the product equal to 1. It is very important to note that $\frac{1}{x}$ is not a generating function. When we do a division, it must always be with a valid generating function (one that has non-zero independent term). It may seem like a simple observation now, but the fact that we can sum, multiply and find inverses of generating functions is what justifies the algebraic manipulations of the following sections. Being able to treat sequences as simple functions is our main goal.

Example 6.1.3 Let $f(x)$ be the generating function associated to the sequence $(1, 1, 1, \ldots)$. Then, $f(x) = \frac{1}{1-x}$.

Solution We have to show that $f(x)(1 - x) = 1$.

$$
\begin{array}{rclclclcl}
f(x) & = & 1 & + & x & + & x^2 & + & \cdots \\
xf(x) & = & & & x & + & x^2 & + & \cdots \\
\hline
(1-x)f(x) & = & 1 & + & 0 \cdot x & + & 0 \cdot x^2 & + & \cdots
\end{array}
$$
□

Exercise 6.1.4 *Let $f(x)$ and $g(x)$ be generating functions associated to the sequences $(a_0, a_1, a_2, \ldots)$ and $(b_0, b_1, b_2, \ldots)$. Show that if k is a positive integer and*

- $b_n = a_{n+k}$ *for all n and $a_t = 0$ if $t < k$, then $f(x) = x^k g(x)$.*
- $b_n = a_0 + a_1 + \cdots + a_n$ *for all n, then $g(x) = \frac{f(x)}{1-x}$.*
- $b_n = t^n$ *for all n, then $g(x) = \frac{1}{1-tx}$.*
- $b_n = n$ *for all n, then $g(x) = \frac{x}{(1-x)^2}$.*
- $b_n = \binom{n}{k}$ *for all n, then $g(x) = \frac{x^k}{(1-x)^{k+1}}$.*
- $b_n = \binom{n+k}{k}$ *for all n, then $g(x) = \frac{1}{(1-x)^{k+1}}$.*
- $b_n = \binom{k}{n}$ *for all n, then $g(x) = (1+x)^k$.*
- $b_n = t^n \binom{k}{n}$ *for all n, then $g(x) = (1+tx)^k$.*

In this example, we are considering k as a non-negative integer, but we can define $\binom{r}{n}$, where r is any real number and n is a positive integer, as

$$
\binom{r}{n} = \frac{(r)(r-1)(r-2)\cdots(r-n+1)}{n!},
$$

$$
\binom{r}{0} = 1.
$$

If r is an integer, the new definition matches what we know as $\binom{r}{n}$ and furthermore, if $b_n = \binom{r}{n}$, where r is a fixed real number, then $g(x) = (1+x)^r$. Showing that, defined this way, the generating functions of $(1+x)^r$ and $(1+x)^s$ satisfy $(1+x)^r(1+x)^s = (1+x)^{r+s}$ can be very tedious. Another way of showing this is true is presented in Sect. 6.5.

With what we know we can already solve some of the previous examples in the book, like 1.4.1. In order to solve it, recall that the problem was equivalent to finding the number of solutions to the equation $n_1 + n_2 + \cdots + n_9 = 9$ with all n_i non-negative integers. Remember that as generating functions, $\frac{1}{1-x} = 1 + x + x^2 + \cdots$.

We are actually looking for the coefficient of x^9 in $\left(\frac{1}{1-x}\right)^9$. To see why this is the coefficient we want, note that when we expand this product, we have to choose a term of the first $\frac{1}{1-x}$ (say x^{n_1}), a term of the second $\frac{1}{1-x}$ (say x^{n_2}), and so on, so that $n_1 + n_2 + \cdots + n_9 = 9$. Note that each solution adds 1 to the coefficient of x^9. Thus this coefficient is exactly the number we are looking for. However, by Exercise 6.1.4, we have that $\left(\frac{1}{1-x}\right)^9$ is the generating function of the sequence $b_n = \binom{n+8}{8}$. Then the number we are looking for is $\binom{17}{8}$.

6.2   Fibonacci Numbers

Fibonacci numbers are defined using the following recursive formula:

$$F_1 = F_2 = 1, \tag{6.3}$$

$$F_{n+2} = F_{n+1} + F_n. \tag{6.4}$$

We want to find a non-recursive formula for these numbers using their generating function. If $f(x)$ is the generating function of the Fibonacci numbers, then

$$f(x) = 0 + F_1 x + F_2 x^2 + F_3 x^3 + \cdots$$

$$
\begin{aligned}
f(x) &= 0 + F_1 x + F_2 x^2 + F_3 x^3 + \cdots \\
x f(x) &= 0 + 0x + F_1 x^2 + F_2 x^3 + \cdots \\
x^2 f(x) &= 0 + 0x + 0x^2 + F_1 x^3 + \cdots
\end{aligned}
$$

$$\left(1 - x - x^2\right) f(x) = 0 + F_1 x + (F_2 - F_1)x^2 + (F_3 - F_2 - F_1)x^3 + \cdots$$

Using the recursive formula (6.4), the terms with exponent greater than or equal to 3 cancel each other. Since $F_2 = F_1 = 1$, we get

$$f(x) = \frac{x}{1 - x - x^2} = \frac{-x}{x^2 + x - 1}.$$

If we consider $\phi = \frac{1+\sqrt{5}}{2}$ (this number is commonly known as the **Golden ratio**), the roots of the polynomial $x^2 + x - 1$ are

$$\frac{1}{\phi} = \frac{-1 + \sqrt{5}}{2}, \qquad -\phi = \frac{-1 - \sqrt{5}}{2}.$$

Then $f(x) = \frac{-x}{(x+\phi)(x-\frac{1}{\phi})}$. In order to simplify this expression, we are going to look for real numbers A and B such that

$$f(x) = \frac{A}{x+\phi} + \frac{B}{x-\frac{1}{\phi}}.$$

That is,

$$\frac{-x}{(x+\phi)(x-\frac{1}{\phi})} = \frac{A}{x+\phi} + \frac{B}{x-\frac{1}{\phi}},$$

$$-x = A\left(x - \frac{1}{\phi}\right) + B(x+\phi),$$

$$-x = (A+B)x + \left(\frac{-A}{\phi} + B\phi\right) = (A+B)x + \left(\frac{B\phi^2 - A}{\phi}\right).$$

Therefore

$$A + B = -1, \qquad B\phi^2 - A = 0,$$

whence

$$A = \frac{-\phi^2}{1+\phi^2} = \frac{-\phi}{\sqrt{5}},$$

$$B = \frac{-1}{1+\phi^2} = \frac{-1}{\phi\sqrt{5}}.$$

It follows that

$$f(x) = \frac{-\phi}{\sqrt{5}}\left(\frac{1}{x+\phi}\right) + \frac{-1}{\phi\sqrt{5}}\left(\frac{1}{x-\frac{1}{\phi}}\right),$$

or

$$f(x) = \frac{-1}{\sqrt{5}}\left(\frac{1}{1-(\frac{-1}{\phi})x}\right) + \frac{1}{\sqrt{5}}\frac{1}{(1-\phi x)}.$$

If we consider the generating functions

$$g(x) = \frac{1}{1-(\frac{-1}{\phi})x}, \qquad h(x) = \frac{1}{1-\phi x},$$

we have that

$$f(x) = \frac{h(x) - g(x)}{\sqrt{5}}.$$

But $g(x)$ is the generating function associated to the sequence $(b_0, b_1, b_2, \ldots)$ where $b_k = (\frac{-1}{\phi})^k$ for all k, and $h(x)$ is the generating function associated with the sequence $(c_0, c_1, c_2, \ldots)$ where $c_k = \phi^k$ for all k. Using this we obtain the following formula for the Fibonacci numbers (know as **Binet's formula**[1]):

$$F_n = \frac{\phi^n - (\frac{-1}{\phi})^n}{\sqrt{5}}.$$

This trick can always be used when we have a recursive formula. We use the recursion to see the generating function as a quotient of polynomials. "Splitting" this fraction into simpler ones is a know technique, called **partial fraction decomposition**. This is a standard technique for manipulation of polynomials. If we have a polynomial $P(x)$ written as a product of terms of the type $(x - t)^n$, we can express $\frac{1}{P(x)}$ as a sum of terms of the type $\frac{A}{(x-t)^m}$ with $m \leq n$. If the polynomial cannot be decomposed this way,[2] then the decomposition can be done up to factors of the type $\frac{1}{(x^2+ax+b)^n}$. In this case we use a sum with terms $\frac{Ax+B}{(x^2+ax+b)^m}$ with $m \leq n$.

The main advantage of using generating functions is that we do not need to know the closed-form formula beforehand. Inductive proofs that Binet's formula does give the Fibonacci numbers give no insight on how the formula was obtained, and thus seem somehow artificial.

6.3   Catalan Numbers

We have seen how generating functions can be used to find closed-form formulas for sequences with a simple recursive formula. However, they can also be used when the recursive formula is more difficult. Let us show this by obtaining a formula for the **Catalan numbers** $(c_0, c_1, c_2, \ldots)$. These numbers are defined as follows

$$c_0 = 1, \tag{6.5}$$

$$c_n = c_{n-1}c_0 + c_{n-2}c_1 + \cdots + c_0 c_{n-1}. \tag{6.6}$$

Note that this recursive formula is similar to the definition of the product of generating functions (6.2). Let $C(x)$ be the generating function of the Catalan numbers. Then,

$$C(x) = c_0 + c_1 x + c_2 x^2 + \cdots,$$

$$C(x)^2 = c_0^2 + (c_0 c_1 + c_1 c_0)x + (c_2 c_0 + c_1 c_1 + c_0 c_2)x^2 + \cdots.$$

---

[1] Although J.P.M. Binet is given the credit for this formula, it has been traced back to Euler in 1765 and de Moivre in 1730. See (for example) **The Art of Computer Programming** by D.E. Knuth for references.

[2] One can always do it by using complex numbers, but this is rarely needed in olympiad problems.

Using the recursive formula (6.6) we obtain

$$C(x)^2 = c_1 + c_2 x + c_3 x^2 + \cdots.$$

Consequently,

$$xC(x)^2 = C(x) - 1,$$

$$x^2 C(x)^2 = xC(x) - x,$$

$$x^2 C(x)^2 - xC(x) = -x,$$

$$x^2 C(x)^2 - xC(x) + \frac{1}{4} = -x + \frac{1}{4},$$

$$\left(xC(x) - \frac{1}{2}\right)^2 = -x + \frac{1}{4} = \frac{1 - 4x}{4}.$$

Note that only one generating function can satisfy this, as the Catalan numbers are defined uniquely by their recursive formula. This means that using the quadratic formula should give us the generating function we want. However, at first glance it seems to give two options. Namely,

$$xC(x) = \frac{1 + (1 - 4x)^{\frac{1}{2}}}{2}, \qquad xC(x) = \frac{1 - (1 - 4x)^{\frac{1}{2}}}{2}.$$

Note that, by Exercise 6.1.4, $(1 - 4x)^{\frac{1}{2}}$ is the generating function of the sequence $(a_0, a_1, a_2, \ldots)$, where $a_k = (-4)^k \binom{\frac{1}{2}}{k}$, so the independent term is 1. Since $xC(x)$ has no independent term, one option is eliminated. So

$$xC(x) = \frac{1 - (1 - 4x)^{\frac{1}{2}}}{2}.$$

Thus, if we calculate the $(k+1)$-th term of the sequence $x C(x)$, we have that

$$
\begin{aligned}
c_k &= -\frac{1}{2}(-4)^{k+1} \binom{\frac{1}{2}}{k+1} \\
&= \frac{(-1)^k}{2} 4^{k+1} \frac{(\frac{1}{2})(\frac{1}{2} - 1)(\frac{1}{2} - 2)\cdots(\frac{1}{2} - k)}{(k+1)!} \\
&= \frac{(-1)^k}{2} 4^{k+1} \frac{(\frac{1}{2})(\frac{-1}{2})(\frac{-3}{2})\cdots(\frac{-(2k-1)}{2})}{(k+1)!} \\
&= \frac{(-1)^k}{2} 4^{k+1} \frac{(-1)(-3)\cdots(-(2k-1))}{2^{k+1}(k+1)!}
\end{aligned}
$$

$$= \frac{4^k}{2^k} \frac{1 \cdot 3 \cdot 5 \cdots (2k-1)}{(k+1)!}$$

$$= \frac{k!2^k \cdot 1 \cdot 3 \cdot 5 \cdots (2k-1)}{k!(k+1)!}.$$

Note that $k!2^k$ is the product of all even numbers between 2 to $2k$. So, we get

$$c_k = \frac{(2k)!}{k!^2(k+1)}$$

$$= \frac{\binom{2k}{k}}{k+1}.$$

We can also prove that the Catalan numbers satisfy $c_n = \frac{\binom{2n}{n}}{n+1}$ without using generating functions (and without knowing beforehand that this is the answer).

To do this, consider the number c'_n of paths in the edges of an $n \times n$ board that start in the lower left corner, end in the top right corner and only move up or to the right and never go above the diagonal joining these two corners (they are allowed to touch it). We define $c'_0 = 1$. These are going to be the "good" paths. (See Fig. 6.1.)

We first show that the sequence $(c'_0, c'_1, c'_2, \ldots)$ has the same recursive formula as the Catalan numbers. For this, label $P_0, P_1, \ldots, P_n$ the points on the diagonal. We know that every path starts in P_0 and ends in P_n. Let us see how many paths there are such that the first time they touch the diagonal again is in P_k. From P_k to P_n there are exactly c'_{n-k} ways to complete the path. Since the path cannot touch the diagonal again from P_0 to P_k, it cannot go above the line parallel to the diagonal but one square down. Thus there are c'_{k-1} ways to complete the path. Since k can run from 1 to n we get

$$c'_n = c'_0 c'_{n-1} + c'_1 c'_{n-2} + \cdots + c'_{n-1} c'_0.$$

Fig. 6.1 Good paths are those that stay in the shaded region

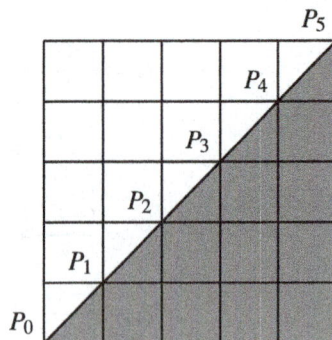

Thus, $c_k = c'_k$ for all k. Now consider all the "bad" paths. That is, all the paths that do go above the diagonal. For each bad path we can consider the first time it goes above the diagonal. Using this point, we can reflect the beginning of the path across a line, as in the figure below.

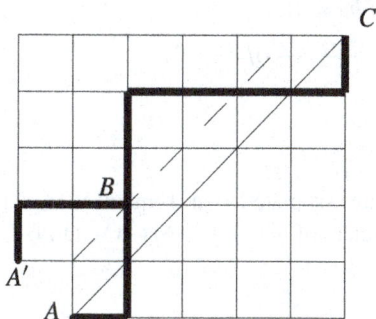

The initial path goes from the point A to the point C. The path after the reflection goes from A' to C. Note that this operation can be inverted. That is, for every path from A' to C we can consider the first time it gets to the parallel of the main diagonal that is one square up and reflect the beginning to obtain a bad path (from A to C). Thus, there are as many bad paths as paths from A' to C. We know that there are $\binom{2n}{n-1}$ paths from A' to C, since this is the number of paths on an $(n-1) \times (n+1)$ board. Also, the total of bad and good paths is $\binom{2n}{n}$. Thus, the number of good paths is $\binom{2n}{n} - \binom{2n}{n-1} = \frac{\binom{2n}{n}}{n+1}$.

6.4   The Derivative

When we work with functions on the real numbers the derivative is a very useful concept. The derivative is basically the slope of a tangent line at a given point to the graph of our function. Its behavior gives us a lot of information about the original function. If we think of generating functions as formal series (i.e. without evaluating at a point), we would like to have a tool similar to the derivative. In this section we define the derivative for generating functions, and in the following one we will relate it to the classic definition of the derivative. The approach in this section is heavily motivated by the usual properties of the derivative, although absolutely no calculus is necessary.

What we want to do is to assign to each generating function $f(x)$ another generating function $f'(x)$ (that is, a sequence) which is called the **derivative** of $f(x)$.

If $f(x)$ is associated with the sequence $(a_0, a_1, a_2, \ldots)$, then $f'(x)$ is the generating function associated with the sequence $(b_0, b_1, b_2, \ldots)$, where $b_n = (n+1)a_{n+1}$ for all n. Those who are familiar with the usual derivatives may notice that this definition is the same for polynomials. Now let us check that with this new definition, we have the usual properties of the derivative.

Proposition 6.4.1 *If $f(x)$ and $g(x)$ are generating functions associated with the sequences $(a_0, a_1, a_2, \ldots)$ and $(b_0, b_1, b_2, \ldots)$, respectively, then*

- $(f + g)'(x) = f'(x) + g'(x)$;
- $(fg)'(x) = f'(x)g(x) + f(x)g'(x)$;
- $(\frac{1}{g})'(x) = \frac{-g'(x)}{g^2(x)}$ *(if $b_0 \neq 0$)*;
- $(\frac{f}{g})'(x) = \frac{f'(x)g(x) - f(x)g'(x)}{g^2(x)}$ *(if $b_0 \neq 0$)*.

Proof

- Left as exercise.
- If $f'(x)$ and $g'(x)$ are associated to the sequences $(c_0, c_1, c_2, \ldots)$ and $(d_0, d_1, d_2, \ldots)$, then the n-th coefficient of $f'(x)g(x) + f(x)g'(x)$ is

$$\sum_{k=0}^{n} c_k b_{n-k} + \sum_{k=0}^{n} a_k d_{n-k}$$

$$= \sum_{k=0}^{n} (k+1)a_{k+1}b_{n-k} + \sum_{k=0}^{n} (n-k+1)a_k b_{n-k+1}.$$

Note that the first sum can start from $k = -1$ and the second one can end in $k = n + 1$ without changing their values, that is, we can write

$$\sum_{k=-1}^{n} (k+1)a_{k+1}b_{n-k} + \sum_{k=0}^{n+1} (n-k+1)a_k b_{n-k+1}.$$

If in the first sum we make the change of variables $t = k + 1$ and in the second one $t = k$, we get

$$\sum_{t=0}^{n+1} t a_t b_{n+1-t} + \sum_{t=0}^{n+1} (n-t+1)a_t b_{n-t+1}$$

$$= \sum_{t=0}^{n+1} (n+1)a_t b_{n+1-t}$$

$$= (n+1) \sum_{t=0}^{n+1} a_t b_{n+1-t},$$

which is precisely the coefficient of the n-th term of $(fg)'(x)$.

- If $h(x) = (\frac{1}{g})(x)$, then $(hg)(x) = 1$. Using the previous result,

$$0 = (hg)'(x) = h'(x)g(x) + h(x)g'(x),$$

$$-h(x)g'(x) = h'(x)g(x),$$

$$-g'(x) = h'(x)g^2(x),$$

as we wanted.

• If we denote $(\frac{1}{g})(x)$ by $h(x)$, then

$$(fh)'(x) = f'(x)h(x) + f(x)h'(x),$$

$$(fh)'(x) = \frac{f'(x)}{g(x)} + \frac{-f(x)g'(x)}{g^2(x)}. \qquad \square$$

Now that we know about the derivative, we can use it to find generating functions in a faster way.

Example 6.4.2 The sum of the first n squares is $\frac{n(n+1)(2n+1)}{6}$.

Solution The generating function of $(1, 1, 1, \ldots)$ is $g(x) = \frac{1}{1-x}$. Then, the generating function of $(0, 1, 2, 3, \ldots)$ is $h(x) = xg'(x) = \frac{x}{(1-x)^2}$. Then, the generating function of $(0, 1^2, 2^2, 3^3, \ldots)$ is $p(x) = xh'(x) = \frac{x(1+x)}{(1-x)^3}$. Thus, the generating function $f(x)$ for the sum of the first n squares is $\frac{x(1+x)}{(1-x)^4}$. But notice that

$$f(x) = \frac{x^2}{(1-x)^4} + \frac{x}{(1-x)^4}.$$

By Exercise 6.1.4, the first term has as n-th coefficient $\binom{n+1}{3}$ and the second term has $\binom{n+2}{3}$. Thus, the sum of the first n squares is $\binom{n+1}{3} + \binom{n+2}{3} = \frac{n(n+1)(2n+1)}{6}$. $\square$

Note that from this solution we can obtain $n^2 = \binom{n+2}{3} - \binom{n}{3}$. This can be used to find formulas for the sum of the first n odd squares directly, without using the formula for the sum of all squares.

6.5   Evaluating Generating Functions

If we have a generating function $f(x)$ of a sequence $(a_0, a_1, \ldots, a_n, \ldots)$ such that $a_k = 0$ for k larger than some k_0, then it is simply a polynomial. It then makes sense to substitute x for any real number and study $f(x)$ as an ordinary function. However, the question of whether we can do this for infinite sequences is more complicated. The theorems that justify the algebraic manipulations of this chapter are included in most good books on differential calculus. The reader that is interested in how these tools can be proved formally should read the sections on convergence of series and Taylor series in such books. Let us see what happens with the generating function of the sequence $(1, 1, 1, \ldots)$. We know that its generating function is

$$1 + x + x^2 + \cdots = \frac{1}{1-x},$$

where by the equality we mean that it is the multiplicative inverse of the generating function $1 - x$. However, if we recall Exercise 1.2.5, we know that for any real number $q \neq 1$ and a positive integer n, we have that

$$1 + q + q^2 + \cdots + q^n = \frac{1 - q^{n+1}}{1 - q}.$$

Note that if $|q| < 1$,[3] then q^{n+1} goes to 0 as n goes to infinity. Thus the sum $1 + q + q^2 + \cdots + q^n$ gets very close to $\frac{1}{1-q}$ as n goes to infinity. Therefore, it make sense to evaluate $f(x)$ in the interval $(-1, 1)$, and in that case the function is equal to $\frac{1}{1-x}$. If this happens, we say that the generating function **converges**. One should note that outside of the interval $(-1, 1)$ we cannot evaluate $f(x)$, as the sum of the first n terms of the generating function does not approach anything as $n \to \infty$.

Now suppose that we have the generating function $f(x)$ of a sequence $(a_0, a_1, \ldots, a_n, \ldots)$, and suppose that there are no 0 terms. We can consider the sequence $(b_0, b_1, b_2, \ldots)$, where $b_i = |a_i|$ for all i, and the fractions

$$\frac{b_1 x}{b_0}, \frac{b_2 x}{b_1}, \frac{b_3 x}{b_2}, \ldots.$$

If, after some point, all these numbers are smaller than a positive constant α, then the generating function $g(x) = b_0 + b_1 x + b_2 x^2 + \cdots$ grows at most as fast as $1 + \alpha + \alpha^2 + \alpha^3 + \cdots$ (up to a constant multiplying everything). So $g(x)$ should converge as long as $\alpha < 1$. Moreover, working with $g(x)$ is enough, as the **absolute convergence theorem** says that if the sum Y of the absolute values of the terms of an infinite sum X converges, then so does X. Putting this together we get

Proposition 6.5.1 *Given a sequence $(a_0, a_1, a_2, \ldots)$ with no zero terms, suppose there is a positive α such that for all $n \geq N$ for some N,*

$$\alpha \leq \frac{|a_n|}{|a_{n+1}|}.$$

Then the generating function $f(x) = a_0 + a_1 x + a_2 x^2 + \cdots$ can be evaluated in the interval $(-\alpha, \alpha)$.

If there are zero terms then this method can run into some difficulties, but trying to compare with a sum of the kind $1 + \alpha + \alpha^2 + \cdots$ is usually a good idea. One should note that there are sequences that grow so fast that their generating function cannot be evaluated at any point other than 0. An example is provided the sequence $(a_0, a_1, a_2, \ldots)$ where $a_n = 2^{n^2}$.

---

[3] By $|y|$ we mean the absolute value of y. That is, $|y| = y$ if y is non-negative, and $|y| = -y$ if y is negative.

One nice thing about evaluating generating functions is that this procedure behaves well with respect to derivation. That is, if we have a generating function $f(x)$ that converges in an interval $(-\alpha, \alpha)$, then its derivative $f'(x)$ (defined as in the previous section), also converges in $(-\alpha, \alpha)$. Moreover, $f'(x)$ is the derivative of $f(x)$ in the classic sense, that is, the rate of change of $f(x)$.

Now suppose that we want to work the other way around. That is, we are given a function f that is infinitely differentiable at 0 and we want to find a sequence $(a_0, a_1, \ldots, a_n, \ldots)$ such that f is the generating function of this sequence. If this is the case, then we must have

$$f(x) = a_0 + a_1 x + a_2 x^2 + \cdots.$$

Evaluating at 0 gives $f(0) = a_0$. If we take the derivative of both sides, we obtain

$$f'(x) = a_1 + 2a_2 x + 3a_3 x^2 + \cdots.$$

Evaluating at 0 gives $f'(0) = a_1$. If we do this again, we obtain

$$f''(x) = 2a_2 + 2 \cdot 3a_3 x + 3 \cdot 4a_4 x^2 + \cdots.$$

Evaluating at 0 gives $\frac{f''(0)}{2} = a_2$. If we continue in this way, then we obtain $a_n = \frac{f^{(n)}(0)}{n!}$, where by $f^{(n)}(x)$ we mean the value of the n-th derivative of f at x. This generating function is called the **Taylor series** of f at 0,

$$f \mapsto \frac{f(0)}{0!} + \frac{f'(0)}{1!}x + \frac{f^{(2)}(0)}{2!}x^2 + \frac{f^{(3)}(0)}{3!}x^3 + \cdots.$$

One should note that not all functions have convergent Taylor series, but in many cases they do and the series is equal to the function.

In particular, this gives a lot more meaning to the extended binomial coefficients. This is because if we consider the function $f(x) = (1+x)^r$, then

$$f^{(n)}(x) = r \cdot (r-1) \cdots (r-n+1)(1+x)^{r-n}.$$

This is why it is natural to think of the coefficients $\binom{r}{n}$ as

$$\binom{r}{n} = \frac{r \cdot (r-1) \cdots (r-n+1)}{n!}.$$

It also explains why their generating function should satisfy the properties we want, such as $(1+x)^r (1+x)^s = (1+x)^{r+s}$, which is not easy to see if these functions are defined only as formal series. The formal arguments needed to show that the generating function $g(x)$ of the coefficients $\binom{r}{n}$ is equal to the function $f(x) = (1+x)^r$ can be followed by the interested reader in the footnote.[4]

Example 6.5.2 Find the Taylor series of $\sin(x)$ and $\cos(x)$.

Solution We know that $\sin(0) = 0, \cos(0) = 1$ and $\sin'(x) = \cos(x)$, $\cos'(x) = -\sin(x)$. Then, finding their n-th derivative only depends only on the value of n modulo 4. This gives

$$\sin(x) = x - \frac{x^3}{3!} + \frac{x^5}{5!} - \frac{x^7}{7!} + \cdots,$$

$$\cos(x) = 1 - \frac{x^2}{2!} + \frac{x^4}{4!} - \frac{x^6}{6!} + \cdots. \qquad \square$$

6.6   Problems

Problems in this section can be solved using generating functions. However, the reader should also try to find alternative solutions without using this technique.

Problem 6.1 Consider $\exp(x) = \frac{1}{0!} + \frac{x}{1!} + \frac{x^2}{2!} + \cdots$.[5] That is, the generating function associated with the sequence $(\frac{1}{0!}, \frac{1}{1!}, \frac{1}{2!}, \ldots)$. Show that

- $\exp'(x) = \exp(x)$;
- $\exp(x + y) = \exp(x)\exp(y)$.

Problem 6.2 Show that $F_n = \binom{n-1}{0} + \binom{n-2}{1} + \binom{n-3}{2} + \cdots$.

Note: When the number in the lower part of the binomial coefficients is bigger than the one on top, the terms of the sum are 0.

Problem 6.3 Find a closed-form formula for the Lucas numbers, defined by $L_1 = 2$, $L_2 = 1$ and the recursive formula $L_{n+2} = L_{n+1} + L_n$.

Problem 6.4 Find a closed-form formula for the sequence $\{a_n\}$ such that $a_0 = 1$, $a_1 = 3$, $a_{n+1} - 3a_n + 2a_{n-1} = 2^n$ ($n \geq 1$).

---

[4]We need to show that these Taylor series converge and behave well. However, the formal arguments needed for this are much simpler than what needs to be done otherwise. The interested reader may want to consider $g(x)$ as this Taylor series and see that it satisfies $\frac{g'(x)}{g(x)} = \frac{r}{1+x}$. Integrating on both sides and using the value of $g(0)$ should give $g(x) = (1+x)^r$ (as a real function). Why are all these steps valid?

[5]Note that $\exp(-1)$ is the same number that seemed to appear in Example 1.4.6.

Problem 6.5 Given the set $A = \{1, 2, 3, \ldots, n\}$, consider for each non-empty subset of A the inverse of the product of its elements. Find the sum of all these numbers.

Problem 6.6 $2n$ points are given on a circle. They are split into n pairs and each pair is joined by a segment. Find the probability[6] that no two of these segments intersect.

Problem 6.7 Consider the number A of sequences of n letters a or b such that for every $1 \leq k \leq n$, among the first k letters there are never more b's than a's. Show that $A = \binom{n}{\lfloor \frac{n}{2} \rfloor}$.

Problem 6.8 Let k be a positive integer. Show that

$$\binom{2k}{k}\binom{0}{0} + \binom{2k-2}{k-1}\binom{2}{1} + \binom{2k-4}{k-2}\binom{4}{2} + \cdots + \binom{0}{0}\binom{2k}{k} = 4^k.$$

Problem 6.9 Let a_n be the number of sequences of size n using only the numbers $\{1, 2, 3, 4\}$, with an odd number of ones. Find a closed-form formula for a_n.

Problem 6.10 Show that the sum of the first n cubes is $[\frac{n(n+1)}{2}]^2$.

Problem 6.11 (Austria 2011) Each brick of a set has 5 holes in a horizontal row. We can either place pins into individual holes or brackets into neighboring holes. No hole is allowed to remain empty. We place n such bricks in a row in order to create patterns running from left to right in which no two brackets are allowed to follow one another, and no three pins may be in a row. How many patterns of bricks can be created?

Note: A bracket must use two holes of the same brick.

Problem 6.12 Show that (as generating functions)

$$\frac{1}{1-x} = (1+x)(1+x^2)(1+x^4)(1+x^8)\cdots.$$

Problem 6.13 (USA 1996) Decide if there is a set F of the integers such that the equation $a + 2b = n$ with $a, b \in F$ has exactly one solution for every positive integer n.

Problem 6.14 Let $\{a_n\}$ be the sequence such that $a_0 = 1$, $a_1 = 2$ and $a_n = 3a_{n-1} + 4a_{n-2}$, if $n \geq 2$. Find a closed-form formula for a_n.

---

[6]The probability of an event is the ratio of the number of favorable cases to the number of total cases.

Problem 6.15 Consider all ordered 4-tuples of positive integers (i, j, k, l) such that $i + j + k + l = 31$. Consider the set P of products of the elements of these 4-tuples and the sum S of the elements of P. Show that S is divisible by 31.

Problem 6.16 Let n be a positive integer and consider $S = \{1, 2, \ldots, n\}$. Find the number of ways to color white or black some of the numbers of S so that there are no two consecutive numbers painted with the same color.

Problem 6.17 (Canada 2008) A tower path in a rectangular board of unit squares is a path made by a sequence of movements parallel to the sides of the board from one unit square to one of its neighbors, in which every movement begins where the last one ended and so that no movement crosses a square that was previously visited by a movement of the path. That is, a tower path does not intersect itself. Let $R(m, n)$ be the number of tower paths in an $m \times n$ board (m rows, n columns) that start in the bottom left corner and end in the top left corner. For example, $R(m, 1) = 1$ for every positive integer m; $R(2, 2) = 2$; $R(3, 2) = 4$; $R(3, 3) = 11$. Find a formula of $R(3, n)$ for every positive integer n.

Problem 6.18 (IMO 1979) In a vertex of a regular octagon there is a frog. The frog can only jump to adjacent vertices. Let a_n be the number of ways in which the frog can get for the first time to the opposite vertex in n jumps. Show that

$$a_{2n-1} = 0, \qquad a_{2n} = \frac{(2 + \sqrt{2})^{n-1} - (2 - \sqrt{2})^{n-1}}{\sqrt{2}}.$$

Partitions

7

7.1 Partitions

This chapter deals with theory that is not frequently encountered in olympiad problems. However, this is a very good example of how the tools we have developed so far can be used. The theory presented here can still be useful when solving problems.

Given a set A, a **partition** of A is a family of non-empty pairwise disjoint subsets of A whose union is A. In what follows we are going to assume that A is a finite set. A partition of A into k parts can be represented as a distribution of the elements of A in k indistinguishable bins (the elements of A are distinguishable), where in every bin there is at least one element. For example, the partitions of $\{1, 2, 3\}$ are $\{\{1\}, \{2\}, \{3\}\}, \{\{1\}, \{2, 3\}\}, \{\{1, 3\}, \{2\}\}, \{\{1, 2\}, \{3\}\}$ and $\{\{1, 2, 3\}\}$.

Given a positive integer n, a **partition of** n is an ordered list of positive integers $(n_1, n_2, \ldots, n_k)$ such that $n_1 \geq n_2 \geq n_3 \geq \cdots \geq n_k$ and $n_1 + n_2 + \cdots + n_k = n$. That is, the partitions of n are the ways to see n as a sum of positive integers regardless of the order. For example, the partitions of 4 are $(4), (3, 1), (2, 2), (2, 1, 1), (1, 1, 1, 1)$. The partitions of n into k parts can be represented as a distribution of n indistinguishable elements in k indistinguishable bins, so that each bin contains at least one element. Partitions of integers are much harder to count than partitions of sets. However, we can find many relations of the following type.

Example 7.1.1 The number of partitions of n with exactly k parts is equal to the number of partitions of $n + \binom{k}{2}$ with exactly k parts, all of them different.

Solution To each partition $(n_1, n_2, \ldots, n_k)$ of n with k parts we are going to assign the sequence $(m_1, m_2, \ldots, m_k)$ so that $m_i = n_i + k - i$ for all i. It is clear that $(m_1, m_2, \ldots, m_k)$ is a partition of $n + (k-1) + (k-2) + \cdots + 1 + 0 = n + \binom{k}{2}$ and that all its elements are different. Also, the function thus defined is bijective, which gives us what we wanted. □

Given a partition $(n_1, n_2, \ldots, n_k)$, we can give it a graphic representation. This is known as the **Ferrer diagram** of the partition. To do this, we place in the plane n_1 squares in a column. Then we place a column of n_2 squares so that the lowest one is to the right of the lowest square of the original n_1 squares, and so on. For example, if $n = 8$, the Ferrer diagram of the partition $(3, 2, 2, 1)$ is:

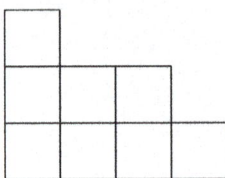

Given a partition $(n_1, n_2, \ldots, n_k)$ we can reflect its Ferrer diagram across the diagonal. What we obtain is the Ferrer diagram of a new partition $(n'_1, n'_2, \ldots, n'_t)$ of n, called the **conjugate partition** of $(n_1, n_2, \ldots, n_k)$. We say that $(n_1, n_2, \ldots, n_k)$ is **self-conjugate** if $(n_1, n_2, \ldots, n_k) = (n'_1, n'_2, \ldots, n'_t)$.

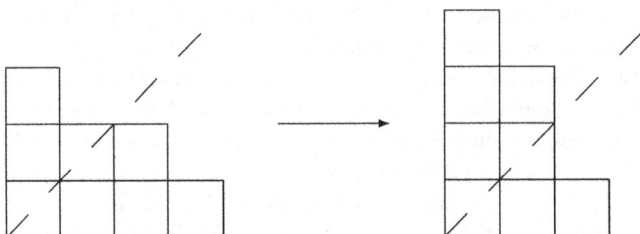

In the figure we show that the conjugate partition to $(3, 2, 2, 1)$ is $(4, 3, 1)$.

Several properties of partitions can be deduced by conjugating Ferrer diagrams.

Example 7.1.2 Show that the number of partitions of n such that the first part has size k is equal to the number of partitions of n with exactly k parts.

Solution This result is almost trivial once we know about conjugated partitions. Given a partition $(n_1, n_2, \ldots, n_t)$ of n, it is clear that the Ferrer diagram of its conjugate has n_1 parts. Since conjugation is a bijective function, we obtain the desired result.   □

7.2   Stirling Numbers of the First Kind

Stirling numbers are two (double) sequences of integers that appear in several counting arguments in combinatorics. As the name suggests, these two sequences are closely related. There are two kinds of Stirling numbers and both can be defined by their combinatorial properties or their algebraic properties (recurrence, generating functions, …). We will define one kind of Stirling numbers by its algebraic properties and the other by its combinatorial properties.

Consider the polynomials $(t)_0 = 1$ and $(t)_n = t(t-1)(t-2)\cdots(t-n+1)$ for $n \geq 1$. When we expand them we denote their coefficients as

$$(t)_n = \sum_{k=0}^{n} s(n,k)t^k. \tag{7.1}$$

The numbers $s(n,k)$ are called the **Stirling numbers of the first kind**. Clearly, $(t)_n$ is the generating function of the sequence $(s(n,0), s(n,1), s(n,2), \ldots)$. Note that the Stirling numbers are not necessarily positive. In fact, if we fix k and increase n by 1 or if we fix n and increase k by 1, the sign of $s(n,k)$ changes (if this is not apparent now, it will be later on). We also have that $s(n,0) = 0$ if $n \geq 1$ and $s(n,n) = 1$ if $n \geq 0$. At first glance, it seems more convenient to define the Stirling numbers of the first kind without these signs. However, when we relate them to the Stirling numbers of the second kind it will be clear why they are used in this way.

Proposition 7.2.1 *The Stirling numbers of the first kind satisfy the recurrence relation $s(n+1,k) = s(n,k-1) - n \cdot s(n,k)$, if $n \geq k \geq 0$.*

Proof Note that

$$\sum_{k=0}^{n+1} s(n+1,k)t^k = (t)_{n+1} = (t-n)(t)_n$$

$$= (t-n)\sum_{k=0}^{n} s(n,k)t^k = \sum_{k=0}^{n} s(n,k)t^{k+1} - \sum_{k=0}^{n} n \cdot s(n,k)t^k$$

$$= \sum_{k=1}^{n+1} s(n,k-1)t^k - \sum_{k=0}^{n} n \cdot s(n,k)t^k.$$

Comparing the coefficients, when $n \geq k \geq 0$, we obtain the desired recurrence relation. If we put $s(n,p) = 0$ if $p < 0$ or $p > n$, the recurrence relation is true regardless of the value of k. $\qquad\square$

If n is a positive integer, denote by $[n]$ the set $\{1, 2, \ldots, n\}$. It turns out that the Stirling numbers of the first kind have a strong relation with the number of permutations of $[n]$ with a fixed number of cycles. We denote by $c(n,k)$ the number of permutations of $[n]$ with exactly k cycles (fixed points count as cycles of length 1).

Proposition 7.2.2 *The numbers $c(n,k)$ satisfy the recurrence relation*

$$c(n+1,k) = c(n,k-1) + n \cdot c(n,k).$$

Moreover, the generating function of the sequence $(c(n,0), c(n,1), c(n,2), \ldots)$ is $c_n(t) = t(t+1)(t+2)\cdots(t+n-1)$.

Proof Consider a permutation σ of $[n+1]$ with exactly k cycles. We have two cases: that $n+1$ is a fixed point and that it is not. If $n+1$ is a fixed point, the rest is a permutation of $[n]$ with exactly $k-1$ cycles, of which there are $c(n, k-1)$. If $n+1$ is not a fixed point, we can remove $n+1$ and send $\sigma^{-1}(n+1)$ to $\sigma(n+1)$ to obtain a permutation of $[n]$ with exactly k cycles. However, any permutation ρ of $[n]$ with exactly k cycles is being counted n times, since we can "add" $n+1$ between j and $\rho(j)$ for all j. In this way we can obtain all the permutations of $[n+1]$ with k cycles with which we counted ρ. Thus in this case there are $n \cdot c(n, k)$ permutations. This gives us the stated recurrence relation.

To show that the generating function is the one given by the proposition, let us use induction over n. It is clear that if $n=1$ the statement is true. If it is true for n, note that

$$
\begin{aligned}
c_{n+1}(t) &= \sum_{k=0}^{n+1} c(n+1, k) t^k \\
&= \sum_{k=0}^{n+1} (c(n, k-1) + n \cdot c(n, k)) t^k \\
&= \sum_{k=0}^{n+1} c(n, k-1) t^k + n \sum_{k=0}^{n+1} c(n, k) t^k \\
&= t c_n(t) + n \cdot c_n(t) = (t + n) c_n(t),
\end{aligned}
$$

and consequently $c_{n+1}(t) = t(t+1) \cdots (t+n)$.

It follows that

$$
\begin{aligned}
\sum_{k=0}^{n} (-1)^{n+k} c(n, k) t^k &= (-1)^n \sum_{k=0}^{n} (-t)^k c(n, k) = (-1)^n c_n(-t) \\
&= (-1)^n (-t)(-t+1) \cdots (-t+n-1) = (t)_n. \qquad \square
\end{aligned}
$$

Thus it is immediate that $s(n, k) = (-1)^{n+k} c(n, k)$. That is, $|s(n, k)|$ is the number of permutations of $[n]$ with exactly k cycles. For this reason the numbers $c(n, k)$ are usually called the **unsigned Stirling numbers of the first kind**.

7.3   Stirling Numbers of the Second Kind

Denote by $S(n, m)$ the number of partitions of $[n]$ into m parts, if $n \geq 1$, and define $S(0, k) = 0$, if $k \geq 1$, and $S(0, 0) = 1$. The $S(n, m)$ are called the **Stirling numbers of the second kind**. These Stirling numbers also satisfy a simple recurrence relation.

Proposition 7.3.1 $S(n+1, k) = S(n, k-1) + k \cdot S(n, k)$.

Proof Consider a partition of $[n+1]$. There are two cases, that $n+1$ is alone in the partition and that it is not. If it is alone, by removing $n+1$, we are left with a partition of $[n]$ into $k-1$ parts, and each of these partitions is only counted once.

Thus there are $S(n, k-1)$ partitions. If $n+1$ is not alone, by removing it we are left with a partition of $[n]$ into k parts. However, we are counting each of these partitions k times. This is because given a partition S of $[n]$ into k parts, we can add $n+1$ to any of its k parts and obtain a partition of $[n+1]$ into k parts that counted S. Thus in this case there are $k \cdot S(n, k)$ partitions. Note that if $n = 0$, the equation still holds. If we also consider $S(n, -1) = 0$ for all n, the recurrence is also true for $k = 0$.

Note that if $n \geq 1$, then $S(n, 0) = 0$ and $S(n, 1) = S(n, n) = 1$. □

Exercise 7.3.2 Show that if $n \geq 2$, then $S(n, 2) = 2^{n-1} - 1$.

The Stirling numbers of the second kind also have a relation with the polynomials $(t)_n$.

Proposition 7.3.3 *If n is a non-negative integer, we have that*

$$t^n = \sum_{k=0}^{n} S(n, k)(t)_k.$$

Proof Let us use induction on n.

If $n = 0$ this is clear. Suppose it is true for n.

Note that

$$t^{n+1} = t(t^n) = t \sum_{k=0}^{n} S(n, k)(t)_k$$

$$= \sum_{k=0}^{n} S(n, k)t(t)_k$$

$$= \sum_{k=0}^{n} S(n, k)\big((t)_{k+1} + k(t)_k\big)$$

$$= \sum_{k=0}^{n} S(n, k)(t)_{k+1} + \sum_{k=0}^{n} k \cdot S(n, k)(t)_k$$

$$= \sum_{k=1}^{n+1} S(n, k-1)(t)_k + \sum_{k=0}^{n} k \cdot S(n, k)(t)_k$$

$$= \sum_{k=0}^{n+1} S(n, k-1)(t)_k + \sum_{k=0}^{n+1} k \cdot S(n, k)(t)_k$$

$$= \sum_{k=0}^{n+1} \big[S(n, k-1) + k \cdot S(n, k)\big](t)_k$$

$$= \sum_{k=0}^{n+1} S(n+1, k)(t)_k.$$ □

Using this we can deduce a relation between the two kinds of Stirling numbers. It is sufficient to notice that

$$t^n = \sum_{k=0}^{n} S(n,k)(t)_k = \sum_{k=0}^{n} \sum_{m=0}^{k} S(n,k)s(k,m)t^m.$$

Then we have that

$$\sum_{k=0}^{n} S(n,k)s(k,m)$$

is 1, if $n = m$, and is 0, if $n \neq m$. With a similar argument we can show that this identity is preserved if we swap the Stirling numbers. That is,

$$\sum_{k=0}^{n} s(n,k)S(k,m),$$

is also 1, if $n = m$, and is 0, if $n \neq m$.

Theorem 7.3.4 *The Stirling numbers of the second kind can be expressed directly by the formula* $S(n,k) = \frac{1}{k!}\sum_{j=0}^{k}(-1)^j \binom{k}{j}(k-j)^n$.

Proof Note that the number of surjective functions from $[n]$ to $[k]$ is $k!S(n,k)$. This is because given any partition of $[n]$ into k parts, we can assign each part to a different element in $[k]$ in $k!$ different ways. Also, the inverse image of the elements of $[k]$ under such a surjective function is a partition of $[n]$ into k parts. Let us count these functions in a constructive way. That is, if we want to produce one of these functions, we have to choose to which element of $[k]$ we send each element of $[n]$. This can be done in k^n ways. However, we have also counted the functions where one element of $[k]$ is not covered, so we have to remove them. To count them, we choose the element of k that is not going to be covered and then send each element of $[n]$ to the other $k-1$ possibilities. This gives $\binom{k}{1}(k-1)^n$ functions. However, if we remove those, we have removed twice each function where two elements of k are not covered, so we have to add them back. Thus, by inclusion-exclusion we obtain the desired formula. □

Exercise 7.3.5 Prove the recursive formula of the Stirling numbers of the second kind using Theorem 7.3.4 exclusively.

7.4   Problems

Problem 7.1 Let l be an integer that is the product of n different primes. Show that the number of ways to write l as product of m positive integers all different from 1 is $S(n,m)$.

Problem 7.2 Show that $(a+b)_n = \sum_{k=0}^{n} \binom{n}{k}(a)_k (b)_{n-k}$.

Problem 7.3 Show that the number of self-conjugate partitions of n is equal to the number of partitions of n where all the parts are odd and different.

Problem 7.4 Show that the number of ways in which a set of mn elements can be decomposed into n parts of m elements each is

$$\frac{(mn)!}{(m!)^n n!}.$$

Problem 7.5 Let $d(n,k)$ be the number of permutations of $[n]$ that do not have fixed points and have exactly k cycles. Show that

- $(-1)^{n+k}s(n,k) = \sum_{t=0}^{n} \binom{n}{t}d(t, t+k-n)$;
- $d(n+1,k) = n \cdot d(n-1, k-1) + n \cdot d(n,k)$.

Problem 7.6 Let $T(n)$ be the number of partitions of $[n]$, if $n \geq 1$, and define $T(0) = 1$. Show that $T(n) = \sum_{k=0}^{n-1} \binom{n-1}{k} T(k)$.

Problem 7.7 Show that if $p(m)$ is the number of partitions of m, then the generating function of the sequence $(p(0), p(1), p(2), \ldots)$ is

$$\prod_{n \geq 1}\left(\frac{1}{1-x^n}\right) = \left(\frac{1}{1-x}\right)\left(\frac{1}{1-x^2}\right)\left(\frac{1}{1-x^3}\right)\cdots.$$

Problem 7.8 Show that the number of partitions of n such that their parts are pairwise different is equal to the number of partitions of n such that all their parts are odd.

Problem 7.9 (Baltic Way 1995) In how many ways can we split the set $\{1, 2, \ldots, 1995\}$ into three pairwise disjoint sets so that none of them contains a pair of consecutive numbers?

Problem 7.10 Show that

$$\binom{n}{m}S(m,m) + \binom{n}{m+1}S(m+1,m) + \cdots + \binom{n}{n}S(n,m) = S(n+1, m+1).$$

Problem 7.11 Given two positive integers n and m, show that

$$\frac{n^{n-1}}{(n-1)!} = \sum_{m=0}^{n} \frac{S(n,m)}{(n-m)!}.$$

Hints for the Problems

8

8.1 Hints for Chap. 1

Hint 1.1 Count the number of ways to choose the rows and columns in which you are going to place the towers.

Hint 1.2 Notice that the number of women that have a man to their left is the same as the number of men who have a woman to their right. How can you count that set?

Hint 1.3 Consider a set of n elements. First choose k of them and paint them red. Then choose r red elements and paint them blue. Then choose s blue elements and paint them green. How does the set end up painted?

Hint 1.4 First choose in which order it is going to put on the shoes and in which order it is going to put on the socks. Then choose the way in which the spider is going to choose a foot to put something on it (either a sock or shoe). Why is this the same as the number of ways to order twice the numbers from 1 to 8 in a list?

Hint 1.5 Paint the first column in any way. How many ways are there to paint the rest of the board?

Hint 1.6 Count the number of lists according to how many of their elements are different from zero.

Hint 1.7 Show that for all i, $\frac{|A_i \cap A_{i+1}|}{|A_i| \cdot |A_{i+1}|} \leq \frac{1}{2}$.

Hint 1.8 Show that for any non-empty even set of columns S there is a row that differs from the top one exactly in S.

P. Soberón, *Problem-Solving Methods in Combinatorics*,
DOI 10.1007/978-3-0348-0597-1_8, © Springer Basel 2013

Hint 1.9 Consider the problem with cards from 1 to n and apply induction on n. Try some small cases to determine how many distributions you want to prove there are.

Hint 1.10 Show that $\sum_{k=0}^{n} k \binom{n}{k}^2 = \sum_{k=0}^{n} (n-k) \binom{n}{k}^2$.

Hint 1.11 Show that if you already have $t_1, t_2, \ldots, t_k$ with $k \leq 99$, then you can choose t_{k+1}.

Hint 1.12 Show that $\binom{m}{s}\binom{s}{r} = \binom{m}{r}\binom{m-r}{m-s}$.

Hint 1.13 Let $2x$ be the sum of all written numbers. Let A be the subset of written numbers with the biggest possible sum that does not exceed x. Show that if it is not possible to label the numbers as the problem asks then, for each r, all the written numbers k with $1 \leq k \leq r$ are in A.

Hint 1.14 Consider the section formed by m consecutive sides and the diagonal that joins the two extremes. Show by strong induction that if $m \leq 100$, then in a triangulation of this section we can find at most $\lfloor \frac{m}{2} \rfloor$ isosceles triangles with two good segments as sides and that if $m > 1003$, then we can find at most $\lceil \frac{m}{2} \rceil$ of these triangles.

8.2   Hints for Chap. 2

Hint 2.1 Show that given any 5 integers, there are 3 of them such that their sum is divisible by 3.

Hint 2.2 How many persons are there to shake their hands with?

Hint 2.3 Show that there are at most n^2 representatives. Construct the seating order by induction.

Hint 2.4 Show that if $a + b > n$ then there is such pair. If $a + b \leq n$, construct a situation where no such pair exists.

Hint 2.5 Use the pigeonhole principle on the pairs (F_k, F_{k+1}) modulo 10^n.

Hint 2.6 Write the numbers as $2^i \cdot j$ with j odd.

Hint 2.7 Number the vertices from 1 to 2007 and consider the sets of 4 of consecutive vertices. How large must k be so that the k vertices always contain one of these sets?

Hint 2.8 Show that there can be no charrúa board with more than $2006 \cdot 2007$ colored squares. To find the number of colorings, focus on the properties of the non-colored squares.

Hint 2.9 Since $r(3,3) = 6$, you already know there is a triangle. What should happen so that there is no other?

Hint 2.10 Count the number pairs of lines of different colors that share a vertex.

Hint 2.11 Consider a person and show that with another 6 he only discussed one topic.

Hint 2.12 Use Proposition 2.2.4.

Hint 2.13 Consider 9 points in the plane and with blue or green segments between them. Show that there is a point with an even number of blue segments (or do Exercise 4.1.2). After that, use arguments similar to the proof of $r(3,3) = 6$.

Hint 2.14 Try to construct such a set of points. What would need to happen for this to be impossible?

Hint 2.15 Count the number of triples (a, b, c) such that a, b, c are in different committees and both a and c know b or neither knows b.

Hint 2.16 Use an argument similar to the one used for Example 2.1.4.

Hint 2.17 Show that if there are $7n + 1$ intervals in a line, then there are $n + 1$ which are pairwise disjoint or 8 of them that have a point in common. Focus on the leftmost point of each interval.

Hint 2.18 How many possible subsets are there? How many possible sums?

Hint 2.19 Paint an arbitrary point x_0 of the second circle in blue (consider the original colored arcs to be red). Place the second circle on top of the first and start to rotate it. If the condition we are looking for is not met, paint in blue on the first circle the point over which x_0 lies.

Hint 2.20 Show that there are at least 51 of these subsets that have more than $\frac{n}{51}$ elements each.

Hint 2.21 Show that there is a student in at least 4 teams. How can those teams be?

Hint 2.22 Solve the problem considering only parallelograms with one side parallel to the x-axis. Why should this be enough?

Hint 2.23 Count the number of pairs of squares of the same row with the same color. How many can there be? Use this to show that $2n^2 - n + 1$ is n-square. To show that with $2n^2 - n$ there can be such a coloring without these 4 squares, consider n boards of size $2n \times (2n - 1)$ such that to go from one to the other we only shift the colors in a cyclic order. How can we paint the first one so that when we paste them together we obtain the desired board?

Hint 2.24 Let a_i and b_i be the number of rows of color i and the number of columns of color i respectively. Why is it that either $a_1 + a_2 + \cdots + a_N > nN$ or $b_1 + b_2 + \cdots + b_N > nN$?

8.3   Hints for Chap. 3

Hint 3.1 Use the chessboard coloring.

Hint 3.2 Show that A can always leave on the table the same number of tokens of each color.

Hint 3.3 Use the parity of the number of coins.

Hint 3.4 Consider the following coloring:

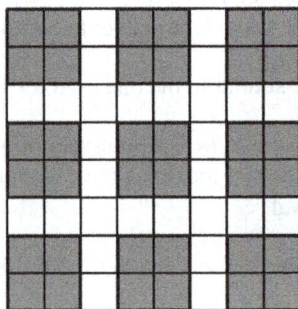

Hint 3.5 Show that in every 5×1 piece there are at least 3 colored squares.

Hint 3.6 Look at the losing positions modulo 11.

Hint 3.7 Seek strategies where only the power of one prime divisor of the last erased number is changed.

Hint 3.8 Count the number of pairs (a, b) of numbers $a \leq b$ that are not in order with respect to the order we want to get. For example, in the first setting $(14, 15)$ is not in order but $(13, 14)$ is.

Hint 3.9 Think of the losing positions modulo 13, and remember that each losing position depends on how many coins were removed in the previous turn.

Hint 3.10 Show that A wins. He starts by removing the card $(1, 997)$.

Hint 3.11 Consider the following coloring of the board:

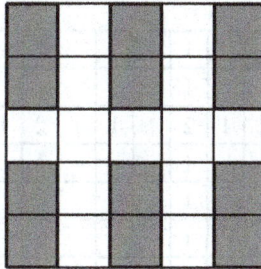

Hint 3.12 If the strategy for B is to increase the minimum difference between any two numbers left and the strategy for A is to reduce the maximum possible difference between two numbers left, what can they guarantee?

Hint 3.13 Consider the largest strip of numbers starting from the far left ordered as B wants them. What can B do if A gives him a number that does not allow him to increase the length of this strip?

Hint 3.14 Stand in the center of the polygon and assign $+1$ or -1 to each vertex depending on what color it has to its right and what color it has to its left.

Hint 3.15 Count the number of pairs (a, C), where C is a 2×2 square and a is a white unit square of C.

Hint 3.16 Assign a 0 to each white square, a 1 to each black square, and a 2 to each green square. What can you say about the number of operations modulo 3?

Hint 3.17 Count the number of piles whose number of coins is a multiple of 4. Search for a strategy based on the parity of this number.

Hint 3.18 Color the rows black and white alternatingly.

Hint 3.19 Show that regardless of which token A removes at first, the rest of the board can be tiled with 2×1 domino tiles. For this it is important that the token A removes is black. Use this tiling to find a winning strategy for B.

Hint 3.20 Use a chessboard coloring on the polygon. Call b the number of black squares and w the number of white squares. Let B be the length of the boundary that is painted black and W the length of the boundary that is painted white. Try to find an equation that relates b, w, B and W.

Hint 3.21 First make sure that there are two S letters separated by two empty spaces so that the game cannot end in a draw. Once you have this, how should B place O's so that A cannot win?

Hint 3.22 First show that the tiles come in pairs. Consider the following coloring to show that at least one of the numbers m or n is divisible by 4:

			1				1		
			1				1		
1	1	1	2	1	1	1	2	1	1
			1				1		
			1				1		
			1				1		
1	1	1	2	1	1	1	2	1	1
			1				1		
			1				1		
			1				1		

Then show how to cover the remaining boards with 3×4 tiles.

Hint 3.23 To show that n is even, color the 5 rows black and white alternatingly. To see that the number of covers is at least $2 \cdot 3^{k-1}$ use strong induction on k. Show that the number of ways to cover the board so that every vertical line cuts at least one tile is at least 2.

8.4   Hints for Chap. 4

Hint 4.1 Consider the longest path in the graph.

Hint 4.2 Consider the smallest connected component.

Hint 4.3 Consider a vertex u of maximal degree.

Hint 4.4 Represent the games as the edges of a graph with 28 vertices. How can assign players to the vertices?

Hint 4.5 Consider a graph with one vertex for each person and an edge if they are seated with exactly one person between them at the table. Color the vertices according to their nationality. What does it mean if two adjacent vertices have the same color?

Hint 4.6 Consider a graph with one vertex per person, a red edge if the persons know each other, and a blue edge if they do not. Consider the largest set of vertices with only red edges.

Hint 4.7 To prove that it is always possible, use induction on n (the number of vertices of G). Do the cases when n is odd and even separately.

Hint 4.8 Consider a graph with one vertex per person and an edge if the persons are friends. Show that there is an odd cycle and let B be the shortest odd cycle. Prove that C, the complement of B, is non-empty and has an edge.

Hint 4.9 Consider two cases, when there is a 4-clique and when there is none.

Hint 4.10 Translate the problem to graphs and use Theorem 4.3.3. Consider the shortest odd cycle.

Hint 4.11 Every non-cyclic set has 3 elements where one of the teams beat the other two. What is the minimum number of those sets? Recall Exercise 1.1.19.

Hint 4.12 Use Hall's Theorem.

Hint 4.13 Use Theorem 4.4.2.

Hint 4.14 Show that you can extend it to an $n \times (k+1)$ board. Consider a graph with $2n$ vertices such that n represent the rows and n represent the numbers from 1 to n. Place an edge between a row and a number if the number is not yet in that row.

Hint 4.15 Draw a graph whose vertices are the points and there is an edge between two of them if in the interior of one of the arcs of the circle there are exactly n vertices. What do you know about this graph?

Hint 4.16 Consider two adjacent vertices v_i and v_j, with degrees $d(v_i)$ and $d(v_j)$. Show that there are at least $d(v_i) + d(v_j) - n$ triangles that use the edge (v_i, v_j). Remember that for any positive numbers $a_1, a_2, \ldots, a_n$ we have that $a_1^2 + a_2^2 + \cdots + a_n^2 \geq \frac{(a_1 + a_2 + \cdots + a_n)^2}{n}$.[1]

Hint 4.17 Show by strong induction that if $6k+2$ teams play 6 rounds there are $k+1$ teams such that no two of them played each other. To do this try to construct the set of $k+1$ teams one at a time, examining all possible cases.

Hint 4.18 Consider a graph G where every vertex represents a city and an edge represents a road. Call t the "radius" of the graph. Consider the subgraph H of G with the same radius, but the smallest number of vertices possible. Show that there are vertices $v_0, v_1, v_2, \ldots, v_t, w$ such that $(v_0, v_1, v_2, \ldots, v_t)$ is the shortest path from v_0 to v_t, to go from v_0 to w at least $t-1$ edges are needed and to go from v_2 to w at least t edges are needed.

---

[1] This is (one of the forms of) the [quadratic mean]-[arithmetic mean] inequality.

8.5   Hints for Chap. 5

Hint 5.1 Use the pigeonhole principle.

Hint 5.2 Count the number of ways to make a committee that has one president and one secretary.

Hint 5.3 Count the number of pairs (P, c), where P is a clown and c is a color that is used by P.

Hint 5.4 For 2003 construct the order. For 2004 show by double counting that this is impossible.

Hint 5.5 Count the pairs (E, p), where E is a student and p is a question that E got right.

Hint 5.6 Count the number T_k of pairs (P, C), where C is a class of k students and P is a pair of students in C.

Hint 5.7 Consider a group of $n - 1$ women and $m + 1$ men. Count, using inclusion-exclusion, how many groups of n persons there are according to the number of men in it.

Hint 5.8 Where does $\alpha \sigma \alpha^{-1}$ send $\alpha(i)$?

Hint 5.9 Suppose there are m associations and let A be the one with the biggest number of members. Count the numbers of associations with at least one member of A in two ways.

Hint 5.10 Notice that the sum is the number of pairs (f, a) where f is a permutation of $\{1, 2, \ldots, n\}$ and a is a number such that $f^{12}(a) = a$.

Hint 5.11 Count the number of pairs (P, t), where P is a permutation of $\{1, 2, \ldots, n\}$ and t is a fixed point of P.

Hint 5.12 Assign a number from 1 to $n + 1$ to each balanced sequence in such a way that every balanced sequence has neighbor sequences with any assigned number.

Hint 5.13 Show that under these assumption every element of B is in exactly two of the A_i. Count the number of pair (a, A_i) such that $a \in A_i$ and a has 0 assigned. For the n that satisfy the condition, use a regular polygon of $2n + 1$ sides to make the assignment.

Hint 5.14 Count the pairs (P, q), where q is a contestant and P is a pair of judges that agree on q (either satisfactory or non-satisfactory).

Hint 5.15 We say that $L = (L_1, L_2, \ldots, L_n)$ is a good list if for every $1 \le i \le n$, L_i is a subset of i elements of $\{1, 2, \ldots, n\}$ and L_i is a subset of L_j for all $i \le j$. Use such lists to do a double counting.

Hint 5.16 Count the number of pairs (P, Q) such that P is a pair of persons and Q is a match in which the persons of P played in different teams. Once you know what the conditions on n should be, form the smallest possible tournament and use that one to form the rest.

Hint 5.17 Assign to each of the q^n colors a sequence of n pearls of q possible colors.

Hint 5.18 Denote by A the set of sequences counted for N and by B the set of sequences counted for M. To each sequence in A assign a sequence of B in the following way: whenever the state of the lamp $n + t$ was changed, instead change the state of lamp t.

Hint 5.19 Count the number of pairs (P, q) such that P is a pair of points of S and q is a point of S in the orthogonal bisector of P.

8.6 Hints for Chap. 6

Hint 6.1 Use Theorem 1.1.8.

Hint 6.2 Consider the generating function $\frac{1}{1-x-x^2}$ and do the substitution $u = x + x^2$.

Hint 6.3 Consider $L_0 = -1$ and use a similar argument to one used in the chapter to find the generating function of the Fibonacci numbers. You should obtain that the generating function of the Lucas numbers is $\frac{-1+3x}{1-x-x^2}$.

Hint 6.4 Show that if f is the generating function of the sequence, then $(1 - 3x + 2x^2)f(x) = 1 - x + x(\frac{1}{1-2x})$.

Hint 6.5 Factorize.

Hint 6.6 Show that the number of ways to join the points by pairs of non-intersecting segments is c_n.

Hint 6.7 Translate te problem to paths in a board and follow an argument similar to the one in the proof of the closed-form formula for Catalan numbers that did not use generating functions.

Hint 6.8 Let $a_t = \binom{2t}{t}$ and let $F(x)$ denote the generating function of the sequence $(a_0, a_1, a_2, \ldots)$. What do you want $F^2(x)$ to be? Do something similar to what was done to find the closed-form formula for the Catalan numbers.

Hint 6.9 Let b_n be the number of sequences of size n with an even number of 1s (with $b_0 = 1$). Let $B(x)$ and $A(x)$ denote the generating functions and of the sequences b_n and a_n show that $A(x) + B(x) = \frac{1}{1-4x}$ and $A(x) = x(B(x) + 3A(x))$.

Hint 6.10 Show that the generating function associated with $(0^3, 1^3, 2^3, 3^3, \ldots)$ is $\frac{x+4x^2+x^3}{(1-x)^4}$.

Hint 6.11 Show that there are 3 kinds of bricks that can be used. Denote by a_n, b_n, c_n the number of sequences that end in the first, second and third kind of bricks, respectively. Then, find recursive equations relating these sequences of numbers.

Hint 6.12 Recall that every number can be written in base 2 in a unique way.

Hint 6.13 Consider the generating function $f_F(x)$ of the set F (the coefficient of x^n has 1 when n is in F and 0 otherwise). Why do we need to establish the relation $f_F(x) f_F(x^2) = \frac{1}{1-x}$ to finish?

Hint 6.14 Show that the generating function for the sequence a_n is $A(x) = \frac{1-x}{1-3x-4x^2}$.

Hint 6.15 Consider $P(x) = x + 2x^2 + 3x^3 + \cdots$. Find the coefficient of the term with exponent 31 in $P(x)^4$.

Hint 6.16 We say that a coloring of S is good if it satisfies the condition of the problem. Let a_n be the number of good colorings where n is painted white, b_n the number of good colorings where n is black, and c_n the number of good colorings where n is not painted. Show that $c_{n+1} = a_n + b_n + c_n$, $a_{n+1} = b_n + c_n$, $b_{n+1} = c_n + a_n$.

Hint 6.17 First notice that the problem is the same as in a $2 \times (n-1)$ board with the path along its edges. Then the path encloses a section to its left. Find a recursive way to describe these sections.

Hint 6.18 Number the vertices of the octagon from 0 to 7 (with the frog starting in 0). Denote by x_n the way the number of ways the frog can be at the vertex 0 after $2n$ jumps without having passed by the vertex 4, y_n as the number of ways to be in the vertex 2 after $2n$ jumps without having passed by the vertex 4 and z_n the number of ways to be in the vertex 6 after $2n$ jumps without having passed by the vertex 4. With this $a_{2n} = y_{n-1} + z_{n-1}$. If $X(x)$, $Y(x)$, $Z(x)$ are the generating functions of the sequences x_n, y_n, z_n, how are they related?

8.7   Hints for Chap. 7

Hint 7.1 Remember the definition of $S(n, m)$.

Hint 7.2 Remember Example 1.1.5.

Hint 7.3 Consider the following assignment:

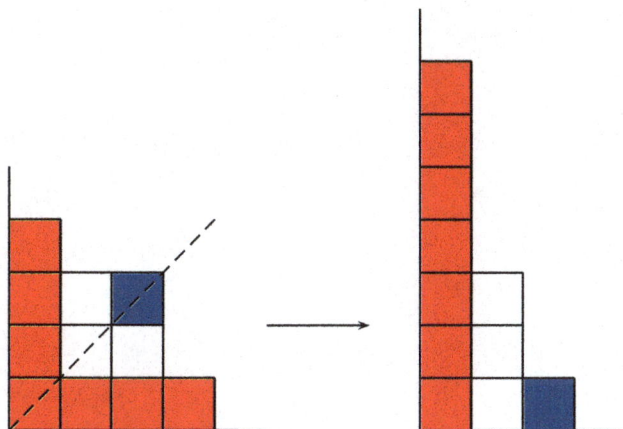

Hint 7.4 Construct the sets of m elements one by one.

Hint 7.5

- Count the number of permutations of $[n]$ with exactly k cycles according to the number of fixed points they have.
- Use an argument similar to the one in Proposition 7.2.2.

Hint 7.6 Decide how many elements will be in the same part as n in the partition.

Hint 7.7 Recall that $\frac{1}{1-x^n} = 1 + x^n + x^{2n} + x^{3n} + \cdots$. Try to assign to each partition of m a way to choose the elements to multiply so that they give a term with exponent m.

Hint 7.8 Follow an argument similar to the one used in Problem 7.7 to find the generating functions of the two sequences of numbers you want to show are equal. Recall that $\frac{1-z^2}{1-z} = 1 + z$.

Hint 7.9 Construct the partition one element at a time.

Hint 7.10 Consider the elements of a partition of $[n + 1]$ into $m + 1$ parts that are not in the same subset as $n + 1$.

Hint 7.11 Given two positive integers n and m, show that

- The number of surjective functions $f : [n] \longrightarrow [m]$ is $m!S(n, m)$.
- The number of functions $f : [n] \longrightarrow [n]$ is n^n.

Solutions to the Problems

9

9.1 Solutions for Chap. 1

Solution 1.1 Note that for the rooks not to attack each other they must use 3 different rows and 3 different columns. There are $\binom{5}{3}^2$ ways to choose them. Then we have to assign to each row its corresponding column where the rooks are going to be placed. There are 3! ways to do this. Thus there are $\binom{5}{3}^2 3! = 600$ ways to place the rooks. □

Solution 1.2 Note that there are 19 women. For each woman that has a woman to its right there is a woman that has a woman to its left. Thus there are 12 women that have a man to their left. Hence, there are 12 men that have a woman to their right. Thus there are 16 men, which gives us a total of 35 persons at the table. □

Solution 1.3 Consider a set of n elements. First we choose k of them and paint them red. Then we choose r red elements and paint them blue. Then we choose s blue elements and paint them green. There are $\binom{n}{k}\binom{k}{r}\binom{r}{s}$ ways to do this. In the end we have s green elements, $r - s$ blue elements and $k - r$ red elements. To paint them this way we can also paint first the s green elements. Then, out the remaining elements we choose $r - s$ and paint them blue, and out of the remaining elements paint $k - r$ in red. There are $\binom{n}{s}\binom{n-s}{r-s}\binom{n-r}{k-r}$ ways to do this. Since we are counting the same colorings, the two results must coincide. □

Solution 1.4 First order the shoes in one row and the socks in another. There are $(8!)^2$ ways to do this. To decide the order in which the spider is going to put on the shoes and socks let us write in a list the numbers from 1 to 8 twice in some order. The spider is going to read this list and, according to the number it reads, put a sock or a shoe on that foot (depending on whether it already has a sock on it or not). To write these number note that there are $(16)!$ to order 16 numbers, but we are counting 2! times each list because the numbers 1 are indistinguishable, 2! times because the numbers 2 are indistinguishable, etc. Thus the number of ways to put the shoes and socks is $\frac{16!(8!)^2}{2^8}$. □

Solution 1.5 Paint the first column in any way. There are 2^8 possibilities. Note that if there are two consecutive squares of the same color, in the next column they must have the colors swapped. Having these two squares with the colors swapped, the whole next column must have opposite colors to the first column. We can go on this way and the whole board coloring is fixed. If there were no two consecutive squares of the same color in the first column then it had to be painted alternating colors. There are only 2 ways to do this. If this happens, then the next column must also be alternating colors, and so on. Thus we only have to choose with which color each column starts. In the first case there were $2^8 - 2$ colorings, while in the second case there were 2^8. Thus there's a total of $2^8 + 2^8 - 2 = 2(2^8 - 1)$ possibilities. □

Solution 1.6 *First Solution* We count the number of lists Alexander wrote. Suppose that there are r elements different from zero in that list. Their absolute values give us a list of positive integers $c_1, c_2, \ldots, c_r$ such that $c_1 + c_2 + \cdots + c_r \leq k$. We can add a positive element c_0 such that $c_0 + c_1 + \cdots + c_r = k + 1$. By Exercise 1.4.3, there are $\binom{k}{r}$ such possible lists. Given any such list, we can choose the r places it will take in Alexander's list and the sign each number will have. That is, it is being counted $\binom{n}{r}2^r$ times. Thus the total number of lists Alexander wrote is

$$\sum_{r=0}^{n} \binom{n}{r}\binom{k}{r}2^r = \sum_{r=0}^{\infty} \binom{n}{r}\binom{k}{r}2^r.$$

The sum may be taken to infinity since if $r > \min\{k, n\}$ the terms are zero. However, written this way it is clear that we get the same number if we swap n and k, as we wanted to prove. □

Second Solution We show how to assign a list of Alexander to a given list Ivan wrote and vice versa. Note that to each list of Alexander we may add one of two possible numbers $a_0, -a_0$ (different from zero) such that $|a_0| + |a_1| + |a_2| + \cdots + |a_n| = k + 1$. The same can be done to the lists of Ivan. If we let $C(n + 1, k + 1)$ denote the number of lists of integers $a_0, a_1, \ldots, a_n$ such that $a_0 \neq 0$ and $|a_0| + |a_1| + |a_2| + \cdots + |a_n| = k + 1$, the problem is reduced to proving that $C(n + 1, k + 1) = C(k + 1, n + 1)$. Call *block* a list such that the first integer is different from zero and the rest are only zeros. Any list counted in $C(n + 1, k + 1)$ may be split into blocks of consecutive elements in a unique way. If x and y are non-zero integers with the same sign, we associate the following pairs of blocks

$$\underbrace{(x, 0, 0, \ldots, 0)}_{|y| \text{ elements}} \longleftrightarrow \underbrace{(y, 0, 0 \ldots, 0)}_{|x| \text{ elements}}.$$

Note that this way every block has exactly one pair. Moreover, doing this to each block of a list counted in $C(n + 1, k + 1)$ gives a unique list counted in $C(k + 1, n + 1)$. Since the operation can be reversed, these sets of lists have the same number of elements, as we wanted to prove. □

Solution 1.7 We show that for all $i \neq j$, $\frac{|A_i \cap A_j|}{|A_i| \cdot |A_j|} \leq \frac{1}{2}$. If A_i and A_j do not intersect, that number is 0. Suppose without loss of generality that $|A_i| \leq |A_j|$. If they do intersect, note that $|A_j| \geq 2$. Also $|A_i \cap A_j| \leq |A_i|$. Thus $\frac{|A_i \cap A_j|}{|A_i| \cdot |A_j|} \leq \frac{1}{2}$. This means that the sum we want is at most n. This value is achieved by letting $A_1 = \{1\}$, $A_2 = \{1, 2\}$, $A_3 = \{2\}$, $A_4 = \{2, 3\}$, ..., $A_{2n-1} = \{n\}$, $A_{2n} = \{n, 1\}$. □

Solution 1.8 Let A be the top row and S be a non-empty set of $2k$ columns. Then S can be partitioned into k pairs $P_1, P_2, \ldots, P_k$ of columns. We can find a sequence $A_1, A_2, \ldots, A_k$ of rows such that A_1 differs from A exactly in P_1, A_2 differs from A_1 exactly in P_2, ..., A_k differs from A_{k-1} exactly in P_k. Thus A_k differs from A exactly in S. Thus, for every non-empty even subset S of columns we can find a row that differs from A exactly in S. If we assign A to the empty set, we have one row for each even subset of columns. By Exercise 1.1.13, this is half the total number of subsets of columns. Thus $n \geq \frac{2^{10}}{2} = 512$.

It is also interesting to know that the number 512 cannot be improved. If we make a table with 10 columns and all the possible $(0, 1)$-rows with an even number of digits 1, it is easy to see that it satisfies the conditions of the problem. □

Solution 1.9 Note that the trick works if and only if two pairs of cards (not from the same box) with the same sum use the same two boxes. Let us consider the problem with cards from 1 to n and prove that the magician can only distribute them in two essentially different ways. The first is to place 1 and n in different boxes and all the other cards in the last box. The second is to distribute them according to their remainder when divided by 3. That is, in a box we place the cards with numbers $1, 4, 7, \ldots$, in another box we have the numbers $2, 5, 8, \ldots$ and in the last box we have the numbers $3, 6, 9, \ldots$. If $n = 3$ this is clearly true. If $n = 4$ this is also easy to see. Suppose it is true for some $n \geq 4$ and we want to prove it for $n + 1$.

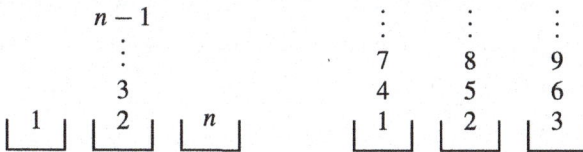

Case 1   case 2

There are two cases. In the first one, in the box where the card $n + 1$ is there is no other card. In the second one, there are other cards in the box with the card $n + 1$. In the second case, we may remove the card $n + 1$ and the trick still works (since no box is empty). By the induction hypothesis, there are only two possible distributions for the cards from 1 to n. Suppose it is the arrangement where 1 and n are alone. Let us see where we could have placed the card $n + 1$. If it is with the card 1, we can choose the pairs $(n, 3)$ and $(n + 1, 2)$. They have the same sum and use different pairs of boxes, so the trick does not work. If $n + 1$ is in the same box as n, we can choose the pairs $(n + 1, 1)$, $(n, 2)$ and again the trick does not work. Finally, if $n + 1$

is with the other cards, we can choose $(n + 1, 1)$ and $(n, 2)$ and the trick does not work.

Thus, the cards from 1 to n must be distributed according to their remainder when divided by 3. If $n + 1$ is in the same box as n, we can choose the pairs $(n + 1, n - 2)$, $(n, n - 1)$. The trick does not work since $n - 2, n - 1, n$ are in different boxes. If $n + 1$ is in the same box as $n - 1$, then we can take the pairs $(n + 1, n - 2)$ and $(n - 1, n)$. Since $n - 2$, $n - 1$, n are in different boxes, the trick does not work. Thus, $n + 1$ must be with $n - 2$, which means the cards were arranged by remainder when divided by 3.

Now we are only missing the first case, when the card $n + 1$ is alone in a box. Consider the pairs $(n + 1, k)$ and $(n, k + 1)$ with $n - 1 \geq k \geq 1$. The first pair has its elements in different boxes because $n + 1$ is alone. The second pair cannot use the same boxes because it would have to use the box with $n + 1$. Thus, if the trick works, n and $k + 1$ must be in the same box. For the last box not to be empty then, the card with 1 must be in it. Thus we have the distribution where 1 and $n + 1$ are alone and the rest of the cards are in the last box. This completes the induction.

It remains to show that these two distributions work. In the distribution where 1 and n are alone, using the card 1 and any from 2 to $n - 1$ we can get the sums from 3 to n. With the cards 1 and n we can get only the sum $n + 1$. With the cards from 2 to $n - 1$ and the card n we can get the sums from $n + 2$ to $2n - 1$. Thus the trick works. For the arrangement with the remainders, the remainder of the sum when divided by 3 depends on the two boxes where the number where chosen. Thus if a sum is repeated, it used the same two boxes. Note that each distribution has 3! ways to be arranged in the boxes, so the magician has 12 possible ways to set the cards in the boxes.

Note that with $n = 3$ the two distributions coincide. However, since we are only interested in the case $n = 100$, this is not a problem.   □

Solution 1.10 First Solution We know that $\binom{n}{k} = \binom{n}{n-k}$ by Exercise 1.1.10. Thus, $\sum_{k=0}^{n} k\binom{n}{k}^2 = \sum_{k=0}^{n} k\binom{n}{n-k}^2$. If on the right-hand side we substitute with $t = n - k$, we get $\sum_{t=0}^{n} (n - t)\binom{n}{t}^2$. Then

$$2\sum_{k=0}^{n} k\binom{n}{k}^2 = \sum_{k=0}^{n} k\binom{n}{k}^2 + \sum_{k=0}^{n} (n - k)\binom{n}{k}^2$$

$$= n\sum_{k=0}^{n} \binom{n}{k}^2 = n\binom{2n}{n}.$$

The last equality comes from Proposition 1.3.1. Since $\frac{1}{2}\binom{2n}{n} = \binom{2n-1}{n-1}$, we are done.   □

Second Solution Given a group of n men and n women, count the number of committees of n persons with a woman as president. A way to do this is first to count how many committees with k women there are. For this there are $\binom{n}{k}$ ways to choose the women and $\binom{n}{n-k}$ ways to choose the men. After this, there are k ways to choose the woman that is going to be president. Since $\binom{n}{k} = \binom{n}{n-k}$, we obtain that the number of committees is equal to $\sum_{k=0}^{n} k \binom{n}{k}^2$.

Another way to count the number of committees is first to choose which woman is going to be president. For this we have n options. Having chosen the president, out of the $2n - 1$ persons left we have to choose the rest of the members. Thus there are a total of $n\binom{2n-1}{n-1}$ possible committees. □

It is worth noting that if we do not know Proposition 1.3.1, by using these two solutions we can obtain a different proof from the one given in the chapter.

Solution 1.11 Let t_1 be any number in S. Suppose that we have $t_1, t_2, \ldots, t_k$ with $k \leq 99$. Let t_{k+1} be any element of S different from all the t_j we have and let us see what should happen so that there are no intersections between A_{k+1} and any A_j with $j \leq k$.

If there is such an intersection, then there must be a z such that $z = x + t_j$ for some x in A, $j \leq k$, and $z = y + t_{k+1}$ for some y in A (that is, z is an element of $A_{k+1} \cap A_j$). This implies $t_{k+1} = x - y + t_j$. Note that x and y must be different, since t_j and t_{k+1} are not equal. Thus there are at most $101 \cdot 100 \cdot k$ ways to write the numbers on the right-hand side. This gives at most $101 \cdot 100 \cdot k$ numbers with which t_{k+1} would be a problem. If we also consider that t_{k+1} cannot be any of $t_1, t_2, \ldots, t_k$, then there are at most $101 \cdot 100 \cdot k + k$ values that t_{k+1} cannot take. Since $k \leq 99$, we have that $101 \cdot 100 \cdot k + k = 10101 \cdot k \leq 101010 \cdot 99 = 999999$. Since S has 1000000 elements, there is always a value for t_{k+1} that causes no problems. □

Solution 1.12 Consider a set with m elements. First choose s and paint them blue. Then choose r blue elements and paint them red. In the end we have r red elements and $s - r$ blue ones. To paint the set this way we can first choose the r elements that are going to be red and then choose out of the other $m - r$ the $m - s$ ones that are not going to be blue. Thus we have $\binom{m}{s}\binom{s}{r} = \binom{m}{r}\binom{m-r}{m-s}$. Using this,

$$\sum_{s=r}^{m} \binom{m}{s}\binom{s}{r}(-1)^{m-s} = \sum_{s=r}^{m} \binom{m}{r}\binom{m-r}{m-s}(-1)^{m-s}$$

$$= \binom{m}{r}\sum_{s=r}^{m}\binom{m-r}{m-s}(-1)^{m-s}.$$

If we substitute here $k = m - s$, then when $s = m$ we have $k = 0$ and when $s = r$ we have $k = m - r$. With this we obtain

$$\binom{m}{r}\sum_{k=0}^{m-r}\binom{m-r}{k}(-1)^k,$$

which is 0, by Exercise 1.1.13. □

Solution 1.13 Let $2x$ be the sum of all written numbers. Let A be a subset of the written numbers with the biggest possible sum that does not exceed x, and let B be the rest of the numbers written. If the sum of the elements in A is x, we label them with $\circ$ and the elements of B with $\times$. Suppose that the sum is not x. That is, the sum is $x - p$ for some positive integer p.

We show by induction on r that A contains all the written numbers k with $1 \leq k \leq r$. For the base of induction, let us see what happens if B has a 1. In that case, we move the 1 to A and we get a new set A' with sum of elements $x - p + 1$. This contradicts the way A was constructed. Now suppose the claim is true for some r and let us see what happens if B has an $r + 1$. Since the number r was written at least once, it must be in A. Then swap an r in A for an $r + 1$ in B. We obtain a set A' that contradicts the maximality of A. We can keep doing this until $r = N$, which means that A has all the written numbers. This is clearly a contradiction. Thus the only possible sum of elements in A is x, as we wanted. □

Solution 1.14 First Solution First note that in the triangles we are counting, the good sides are the ones that are equal. Consider a section formed by m consecutive sides and the diagonal that joins its extremities. We prove by strong induction that the maximum number of isosceles triangles with two good segments as sides that can be formed in a triangulation of this section is $\lfloor \frac{m}{2} \rfloor$ if $m \leq 1003$ and $\lceil \frac{m}{2} \rceil$ if $m > 1003$.

Suppose $m \leq 1003$. Label the vertices of the section as $p_0, p_1, p_2, \ldots, p_m$. Note that when we triangulate this section, the base of the region (the segment $p_0 p_m$) must be in a triangle. Suppose that this triangle is formed using the vertex p_k. Then we get the triangle $\triangle p_0 p_k p_m$ and the regions with the vertices from p_0 to p_k and from p_k to p_m. If $\triangle p_0 p_k p_m$ is not one of the triangles we are counting, then by induction there are at most $\lfloor \frac{k}{2} \rfloor + \lfloor \frac{m-k}{2} \rfloor$ good triangles. This is equal to $\lfloor \frac{m}{2} \rfloor$ as long as k and $m - k$ are not both odd (in which case it is less). (See Fig. 9.1.)

If $\triangle p_0 p_k p_m$ is good, this means that $k = m - k$ and both are odd numbers (with $m = 4r + 2$ for some r). Thus there are at most $1 + \lfloor \frac{k}{2} \rfloor + \lfloor \frac{m-k}{2} \rfloor$ good triangles. Since k and $m - k$ are both odd, this is precisely $\lfloor \frac{m}{2} \rfloor$.

Fig. 9.1 Example of a region we are considering

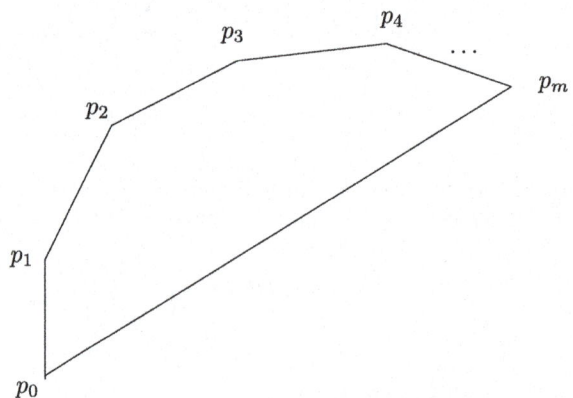

If $m > 1003$, we are going to prove that there are at most $\lceil \frac{m}{2} \rceil$ good triangles in a triangulation of that section (that is, one more triangle if m is odd). When m is odd then $\triangle p_0 p_k p_m$ is good if one of k or $m - k$ is equal to $2006 - m$ and the other is $2m - 2006$. Note that the other odd side ($2006 - m$) is smaller than 1003, so there are at most $1 + \lfloor \frac{2006-m}{2} \rfloor + \frac{2m-2006}{2} = \lceil \frac{m}{2} \rceil$ good triangles.

When m is even, if $\triangle p_0 p_k p_m$ is good then necessarily $k = m - k$ and both numbers are odd. If this is the case, then they are both smaller than or equal to 1003 and we have the same computation as before. If $\triangle p_0 p_k p_m$ is not good and k and $m - k$ are both odd, then at least one of them is smaller than or equal to 1003 (suppose it is k). Thus there are at most $\lfloor \frac{k}{2} \rfloor + \lceil \frac{m-k}{2} \rceil = \frac{m}{2}$ good triangles. This finishes the induction. It is worth noting that in each case the maximum value could be obtained.

Now to triangulate P we must first draw a diagonal l. This diagonal divides the polygon into two sections as before, one of size m at most 1003 and the other of size $2006 - m$. Thus there are at most $\lfloor \frac{m}{2} \rfloor + \lceil \frac{2006-m}{2} \rceil = 1003$ good triangles. Note that this number can always be attained as long as $m \neq 1003$ (since the bounds in the induction could always be achieved). Thus the desired number is 1003. □

Second Solution (with graph theory) For this solution it is highly recommended to have read Sect. 4.2. Consider a graph G with one vertex for each of the 2004 triangles and an edge between two vertices if their triangles share a non-good side. Note that each vertex can have degree 1 or 3. Good triangles have degree 1 (although vertices of degree 1 may not be isosceles!). Note that this graph cannot have a cycle, since removing any edge e makes the graph disconnected (the diagonal represented by e splits all triangles into those that are on one side and those that are on the other, and there are no edges between these two groups).

Let $N_1, N_2, \ldots, N_k$ be the connected components of G (each of them is a tree). If N_i has n_i vertices, suppose x are of degree 1 and y are of degree 3. Counting the number of vertices and the number of edges gives

$$x + y = n_i,$$
$$x + 3y = 2(n_i - 1),$$

which yields $x = \frac{n_i+2}{2}$. This means that the total number of vertices of degree 1 is equal to $\frac{n_1+n_2+\cdots+n_k}{2} + k = 1002 + k$. Note that the vertices of N_i represent triangles that form a polygon, and that the union of the polygons formed by all the connected components is the original 2006-gon. Consider a new graph H with one vertex for each connected component and one edge between two vertices if the corresponding polygons share a side. Note that H is connected. Thus it has at least $k - 1$ edges. If $\triangle_1$ and $\triangle_2$ are triangles in different connected components that share a side, this side must be of odd length (as there is no edge between them!). This implies that $\triangle_1$ and $\triangle_2$ are represented in G by vertices of degree 1 (the side they share must be good, since they are in different connected components of G). Moreover, it is easy to see that at most one of them is a good triangle. Thus at least $k - 1$

Fig. 9.2 Example of a region
we are considering

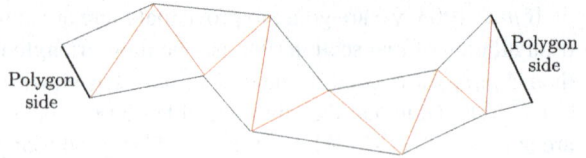

vertices of degree 1 are not good triangles. This implies that there are at most 1003
good triangles. Obtaining a triangulation with 1003 good triangles can be done by
considering 1003 disjoint diagonals that leave exactly 1 vertex on one side and then
1000 other diagonals to complete the triangulation.   □

Third Solution Erase all good segments except the polygon's sides. For any triangle
with a good segment as side, consider the region it is in after removing the good
segments. Note that all triangles had an even number of good segments as sides. This
means that we are forming strips of triangles such that any two consecutive ones
shared good segments. These strips cannot form loops, as any diagonal d divides
the polygon into two parts and the only way to go from a triangle on one side to
the other is through d. Thus each of these regions must end in sides of the polygon.
(See Fig. 9.2.)

Note that in each of these strips there can be at most one good triangle. This
is because a good triangle divides the polygon into 4 parts: the triangle itself, the
section bounded by the even side and two sections x_1, x_2 of equal size bounded by
the odd sides. If there were two good triangles in a strip, one with equal regions
x_1, x_2 and the other with equal regions y_1, y_2, we would have that (under a proper
labeling of the regions) x_1 is a proper subset of y_1 and y_2 is a proper subset of x_2.
This is impossible if both x_1 and x_2 have the same size and y_1 and y_2 have the same
size. Thus, to every good triangle we can assign two sides of the original polygon,
and these are not repeated for any other good triangle. This implies that the number
of good triangles is at most 1003. Finding a construction with 1003 good triangles
can be done as in the previous solution.   □

9.2   Solutions for Chap. 2

Solution 2.1 Given any 5 integers we are going to prove that there are 3 of them
whose sum is divisible by 3. If there are 3 of them congruent modulo 3 to 0, 1
and 2, respectively, then their sum is divisible by 3. If at most two congruences are
used, by the pigeonhole principle one was used at least 3 times. The sum of these 3
numbers is divisible by 3.

Note that given 13 points with integer coordinates, the pigeonhole principle en-
sures that 5 of them have the same first coordinate modulo 3. Of these 5 points, the
second coordinate of 3 of them must have a sum divisible by 3. These are the 3
points we were looking for.

It should be noted that the number 13 is not optimal for this problem. Finding the
smallest number with this property should be a good problem for the reader.   □

Solution 2.2 A person can shake the hands of k persons for any $0 \leq k \leq n - 1$. However, if a person has shaken hands with $n - 1$ others, then no person can shake the hand of 0 others. Thus, there are really only $n - 1$ options: either $\{0, 1, \ldots, n-2\}$ or $\{1, 2, \ldots, n - 1\}$. By the pigeonhole principle, we are done. $\square$

Solution 2.3 We are going to show that there are at most n^2 representatives. Consider the set of pairs (r_i, r_j) such that there is a representative of the country j to the right of a representative of the country i. Note that to any two consecutive representatives we are assigning one of these pairs. If there were at least $n^2 + 1$ representatives, two consecutive pairs of them would have been assigned the same pair, which is impossible.

If $n = 2$ it is clear that we can arrange 4 representatives. If we can seat n^2 representatives of n countries, then each pair (r_i, r_j) with $1 \leq i, j \leq n$ is being assigned exactly once. To add a country, where we assigned the pair (r_i, r_i) we place representatives to have (r_i, r_{n+1}, r_i, r_i) if $i \leq n - 1$. We are also going to change (r_n, r_n) by $(r_n, r_{n+1}, r_{n+1}, r_n, r_n)$. Then every pair (r_i, r_j) with $1 \leq i, j \leq n + 1$ appears exactly once, so we have the seating arrangement for $n + 1$ countries using $(n + 1)^2$ representatives. $\square$

Solution 2.4 Consider the pairs (r, s) where r is a boy and s is a girl that r likes. The are at least a pairs for each boy, so there are at least $a \cdot n$ pairs. Thus, there is a girl that is in at least a of these pairs. That is, there is a girl that at least a boys like. If that girl likes more than $n - a$ boys, then we can find the desired pair. That is, if $a + b > n$ we can find such a pair.

If $a + b \leq n$ we are going to show that this is not necessarily true. For this we are going to number both the girls and the boys from 1 to n. Given a pair (i, j) where i is the number of a girl and j the number of a boy, we say that i likes j if $i + j$ is congruent modulo n to some of $1, 2, 3, \ldots, b$. We say that j likes i if $i + j$ congruent modulo n to some of $n, n - 1, \ldots, n - a + 1$. With this we have the condition of the problem and no pair likes each other. $\square$

Solution 2.5 Consider the pairs (F_k, F_{k+1}) modulo 10^n. There can only be 10^{2n} different ones, so among the first $10^{2n} + 1$ there must be two which coincide. Suppose they are (F_i, F_{i+1}) and (F_j, F_{j+1}) with $i < j$. By the recursive formula we have that (F_{i-1}, F_i) and (F_{j-1}, F_j) must also coincide. We can repeat this argument until the first one is (F_1, F_2) and the second one is (F_s, F_{s+1}) with $s > 1$. Thus F_s and F_{s+1} are both congruent to 1 modulo 10^n. This means that F_{s-1} satisfies the condition we wanted.

It should be noted that we did not use any property of 10^n. If we repeat this argument we can prove that, given any non-zero integer a, there is a Fibonacci number divisible by a. $\square$

Solution 2.6 If we write the numbers from 1 to $2n$ as $2^i \cdot j$ with j odd, j must be between 1 and $2n$. There are only n possibilities for j. Thus, by the pigeonhole principle, there must be two elements from the subset with the same j. It is clear that the biggest of these numbers is divisible by the other.

Note that $n + 1$ is also the smallest number with this property. The set $\{n + 1, n + 2, \ldots, 2n\}$ has n elements and none is divisible by another element in that set. $\square$

Solution 2.7 Number the vertices of the polygon from 1 to 2007. What we want is that any set of k vertices contains at least 4 consecutive ones. Consider the 4-tuples of consecutive vertices. Each vertex uses 4 of these 4-tuples. Thus if $4k > 3 \cdot 2007$ there is a 4-tuple with 4 of the k vertices. That is, if $k \geq 1506$ the condition we want is met. To see that 1506 is the minimum, it is enough to take all the vertices, except the multiples of 4 and 2007. That way we get 1505 vertices with no 4 consecutive ones. $\square$

Solution 2.8 Note that if there are at least $2006 \cdot 2007 + 1$ colored boards, then, as there are 2007 rows, there is a row with at least $2006 + 1$ colored squares. Thus the board is not charrúa. This gives $k \leq 2006 \cdot 2007$.

Notice that if we want a charrúa board with $2006 \cdot 2007$ painted squares, since every row can have at most 2006 painted squares, every row must have exactly 2006 painted squares.

If we focus on the non-colored squares, there must be exactly one in each row and exactly one in each column. If we assign to each row the number of the column in which its non-colored square is, we are assigning all number from 1 to 2007. Thus the number of charrúa boards is equal to the number of permutations of $\{1, 2, \ldots, 2007\}$, which is 2007!. $\square$

Solution 2.9 First Solution Since $r(3, 3) = 6$, we know that there is a monochromatic triangle (suppose it is blue) with vertices v_1, v_2, v_3. Let u_1, u_2, u_3 be the other points and suppose there is no other monochromatic triangle. From u_i (for any $1 \leq i \leq 3$) we cannot have two blue segments to the set $\{v_1, v_2, v_3\}$, so we must have at least two green ones. Thus, from u_1 and u_2 we have at least 4 green segment to $\{v_1, v_2, v_3\}$. At least 2 of them must go to the same v_i, so the segment between u_1 and u_2 cannot be green. The same way the segments between u_1 and u_3 and u_2 and u_3 must be blue. Thus $\triangle u_1 u_2 u_3$ is monochromatic and we are done. $\square$

Second Solution Note that if a triangle is not monochromatic, it must have exactly two angles that use two colors. Each vertex has 5 edges. If they all are from the same color, there are no bi-chromatic angles. If there are 4 from 1 color and 1 from the other, there are 4 bi-chromatic angles. If there are three from one color and two from the other, there are 6 bichromatic angles. With this, there are at most 36 bichromatic angles. Thus, there are at most 18 non-monochromatic triangles. Since there are $\binom{6}{3} = 20$ triangles, at least two of them are monochromatic. $\square$

Solution 2.10 We will show that there are at least 8 monochromatic triangles. Each vertex is part of 7 lines. The largest possible number of pairs of lines of different colors that use this vertex is $4 \cdot 3 = 12$. Hence, the number of pairs of lines of different color that share a vertex is at most $8 \cdot 12$. Each triangle that is not monochromatic uses exactly two of these pairs, so there are at most $8 \cdot 6 = 48$ non-monochromatic triangles. Thus the number of monochromatic triangles is at least $\binom{8}{3} - 48 = 8$.

Note: The problem asks to prove that there are at least 7 because this way it is possible to find a constructive solution without too many difficulties. However, since that solution is heavily case-based and long it will not be presented here.   □

Solution 2.11 Let A be one of the persons. The rest are $16 = 3 \cdot 5 + 1$ persons, and with each one of them he only discussed one of three possible topics. By the pigeonhole principle, there are at least 6 of them with which he only discussed one topic. If any two of those persons discussed that topic, we are done. If not, it means that between those 6 persons only the two other topics were discussed. Since $r(3, 3) = 6$, we can find among them the three persons we were looking for.   □

Solution 2.12 We show this by induction on $l + s$. If $l + s = 2$, then $l = 1, s = 1$ and the statement is true. If it is true for $l + s - 1$, note that, by Proposition 2.2.4, $r(l, s) \leq r(l - 1, s) + r(l, s - 1)$. By the induction hypothesis, $r(l - 1, s) \leq \binom{l+1-3}{l-2}$ and $r(l, s - 1) \leq \binom{l+s-3}{l-1}$. Thus $r(l, s) \leq \binom{l+s-3}{l-2} + \binom{l+s-3}{l-1} = \binom{l+s-2}{l-1}$. The last equality comes from Pascal's formula (Exercise 1.1.11).   □

Solution 2.13 We first show that $r(3, 4) \leq 9$. If at every point there is an odd number of blue segments, then the number of pairs (L, p), where p is a point and L is a blue segment with p as one of its endpoints, is odd. However, each segment is in two pairs, so the number of pairs should be even, which is a contradiction. Thus there is a point P_0 with an even number of blue segments. If it has at least 6 blue segments, we know that among the other endpoints of those segments there is a monochromatic triangle. If the triangle is green we are done, if the triangle is blue we can add P_0 and we are done. If at most 4 blue segments come from P_0, then at least 4 green segments come from P_0. If among the other endpoints of these green segments there is another green segment, we can add P_0 to its endpoints and we are done. If not, those 4 points only have blue segments and, again, we are done.

To show that $r(3, 4) \geq 9$ it suffices to show 8 points with their segments colored either blue or green so that there is no green triangle or 4 points with only blue segments. In the following figure we show 8 points and where the green segments would go (the central point of intersection is not one of the vertices).[1] Clearly no triangle is formed by them. If 4 vertices are picked, there are either two consecutive ones or two opposite ones, so they have at least one green segment. This shows that $r(3, 4) > 8$.

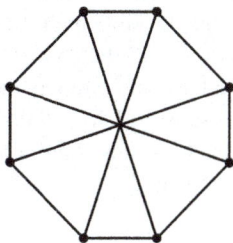

    □

---

[1] This graph is commonly called the **Wagner graph**.

Solution 2.14 Let v_0 be any point. Since an infinite number of segments come from v_0, by the infinite pigeonhole principle there is an infinite number of green segments or an infinite number of blue segments from v_0. Suppose there is an infinite number of blue segments and let A_0 be the set of the other endpoints of these segments. If in A_0 there is a point with an infinite number of blue segments to points of A_0, call it v_1 and define A_1 as the set of all other endpoints of such blue segments. If we can iterate this process indefinitely, then we get a sequence of points $v_0, v_1, v_2, \ldots$ and they are all joined by blue segments.

If we cannot construct some v_{k+1}, that means that in A_k all vertices are joined with blue segments only to a finite number of points in A_{k+1}. Let u_0 be vertex of A_k and repeat the process with the green segments (constructing the sequences u_r and V_r). If again there is a u_{r+1} that we cannot construct, then there is a V_r such that all the points in V_r are connected with green segments only to a finite number of points in V_r. Since V_r is a subset of A_k, the same happens for blue segments. Thus V_r should be finite, but it is infinite by construction. Thus we can always find at least one of these two sequences.   □

Solution 2.15 Consider the triples (a, b, c) such that a, b, c are in different committees and either both a and c know b or neither knows b. We consider the triples (a, b, c) and (c, b, a) as identical. Note that if we pick one congressman from each committee, they form at least one of these triples. Thus the number of triples is at least 100^3. Each triple has its "central" person from some committee, so there are at least $\frac{100^3}{3}$ of these triples with the central person from the same committee. Each of these triples uses one pair of persons, one of each of the other two committees, of which there are 100^2 possible choices. Thus there are at least $\frac{100}{3}$ of these triples that use the same pair of these two committees. One triple can be one where the central person knows the other two or one where he does not know any of the other two. Thus we have at least $\frac{100}{6}$ triples of the same type and with the same "exterior" pair. Since $\frac{100}{6} > 16$, this means there are at least 17 such pairs. The common exterior pair consists of the two congressmen we want, and the 17 triples correspond to the persons that know both of them or neither of them.   □

Solution 2.16 There are 9 primes smaller than or equal to 23. To each of the 1985 numbers we can associate an ordered list of 9 numbers with the power of these 9 primes in its factorization. What we want to see is that there are 4 list so that, when added,[2] they give a list of numbers divisible by 4. First consider the lists modulo 2. There are 512 possible lists. Of every 513 there must be two that are equal modulo 2. Consider any 513 of the lists we have and remove one such pair. We can repeat this process until we are left with at most 511 lists. This means that we have put aside 737 disjoint pairs of lists. For each of these pairs consider its sum (which is a list of only even numbers). These sums modulo 4 only have 512 possibilities. Since we

---

[2]The sum of two lists $(x_1, x_2, \ldots, x_k)$ and $(y_1, y_2, \ldots, y_k)$ with the same number of elements is defined as the list $(x_1 + y_1, x_2 + y_2, \ldots, x_k + y_k)$.

had 737 new lists, there must be two that are equal modulo 4. The 4 integers that gave the lists involved in this new pair are the ones we where looking for. Actually, using this argument we can find at least 112 disjoint 4-tuples of numbers satisfying the condition of the problem. $\square$

Solution 2.17 We show by induction on n that if we are given $7n + 1$ intervals in a line, then there are $n + 1$ of them which are pairwise disjoint or 8 of them that have a point in common. If $n = 0$ the result is clear (as there is only one it is pairwise disjoint). Suppose that the result is true for $n - 1$. Let I_0 be the interval whose leftmost point, P_0, is the one farthest to the right. Any interval that intersects I_0 contains P_0. If at least 7 intervals intersect I_0, then those 7 and I_0 have a point in common. If not, then at least $7(n - 1) + 1$ intervals do not intersect I_0. If there are 8 of them that have a point in common we are done. If not, by the induction hypothesis there are at least n of them which are pairwise disjoint. Adding I_0 to them, we are done.[3] $\square$

Solution 2.18 There are $2^{10} - 1$ non-empty subsets of the 10 numbers. The possible sums go from 10 to $90 + 91 + \cdots + 99 = 945$. By the pigeonhole principle, there are two different subsets A and B with the same sum of elements. If we remove their intersection from both of them, we obtain two disjoint subsets like the ones we wanted. $\square$

Solution 2.19 We are going to paint an arbitrary point x_0 of the second circle in blue. Place the second circle on top of the first and start to rotate it. If the condition we look for is not met, in the first circle paint in blue the point over which x_0 lies. If a point in the first circle was painted blue, it means that one of the original points was in a red arc. Number the original points from 1 to 420. If the point i lies over a red arc, we are coloring some point in blue. Since the sum of the lengths of the arcs is less than 1, the point i is coloring some arcs that measure less than 1 in length. Since this happens for every i, we are coloring some arcs which add up less than 420 in length. Thus there are points on the first circle that are not painted in blue, which is what we wanted to prove. $\square$

Solution 2.20 We show that there are at least 51 of these subsets with more than $\frac{n}{51}$ elements each. Indeed, assuming this is not the case, at least 50 of them have each at most $\frac{n}{51}$ elements each. Their union would not have more than $\frac{50n}{51}$, which is a contradiction. Since there are at least 51 of these sets, by the pigeonhole principle at least two of them have non-empty intersection (call them B and C). Consider the triple B, C, A_i for any A_i different from B and C. If the conclusion of the problem is not true, every A_i must have empty intersection with at least one of B and C. Since there are 99 possibilities for A_i, at least 50 of them must have empty intersection with B or 50 of them must have empty intersection with C. Suppose

---

[3]The number 7 is not important here. In general, given any $nm + 1$ intervals in a line, there are either $m + 1$ of them with a point in common or $n + 1$ of them that are pairwise disjoint.

that it is with B. Since B has more than $\frac{n}{51}$ elements, the union of the other 50 must have less than $\frac{50n}{51}$, contradicting the hypothesis. Thus, at least one triple B, C, A_i works.   □

Solution 2.21 We will show that you cannot have a group of n students with more than n teams in such a way that no two teams have exactly one student in common. For this suppose, on the contrary, that it is possible, and let n be the smallest number of students for which this is possible. Let m be the number of teams ($m > n$). Each team uses 3 students, so the number of pairs (E, a), where E is a team and a is a student in A, is $m \cdot 3 > n \cdot 3$. Thus, there must be a student a in at least 4 teams. Let (a, b, c) be one of such teams. Since a is in another team, that team must have at least one of b and c, it must be of the form (a, b, d). If a is in a team where b is not, it must have both d and c. That is, it must be (a, c, d). But then a cannot be in another other team, so he cannot be in 4 teams or more. Thus, in every team with a we have b. This means that b is in at least 4 teams and (repeating the argument) in every team that b is a must be as well. Let x be a student in any team of the form (a, b, x). Then x cannot be in any other team. If he was, one of a or b should also be there, and then both a and b should be there. Suppose that r students different from a and b are in a team of the type (a, b, r). Note that if we remove a, b and those r students we just mentioned, we are not affecting any team involving the rest of the students. We are left with $n - r - 2$ students who make $m - r$ teams. Since $m > n$, we get that $m - r > n - r - 2$. This means that $n - r - 2$ satisfies what we wanted, which contradicts the minimality of n.

Note that if we only want 32 teams, the condition can be achieved. For this, split the students into 8 groups of 4 each, and in each group consider all the possible triples. We have 32 triples and any two of them have either 0 or 2 students in common. This type of arrangement can only be done if the number of students is a multiple of 4.   □

Solution 2.22 We will show that the largest size is 4021. Consider the subset S_0 of all points (x, y) such that at least one of x and y is equal to 2011. For any 4 points of S_0 either 3 of them lie on the same line or they define a quadrilateral with two orthogonal opposite sides. In either case they do not form a proper parallelogram. Thus S_0 has 4021 points and is parallelogram-free.

Suppose that S is a subset of at least 4022 points. We show that S contains a parallelogram with two sides parallel to the x-axis. For this, given two points of S on the same row of G, consider their distance. If there are two pairs on different rows with the same distance, they form a proper parallelogram. Assign to each row the different distances you can find among its points. Note that if there are m points on the same row, they define at least $m - 1$ different distances.

Let $m_1, m_2, \ldots, m_{2011}$ the number of points on each row. The number of assigned distances is at least

$$\sum_{i=1}^{2011}(m_i - 1) = \left(\sum_{i=1}^{2011} m_i\right) - 2011 \geq 2011.$$

However, there are only 2010 possible distances to assign. By the pigeonhole principle, there is at least one distance that was assigned twice, as we wanted to prove. □

Solution 2.23 Consider the number of pairs of squares in the same row with the same color. Each row has at least n of these pairs, since there are at most n squares of different colors and each new square forms at least one additional pair.[4] Thus a $2n \times k$ board has at least kn of these pairs.

However, these pairs can be any of the $\binom{2n}{2}$ pairs of squares in a row with any of the n colors, so there are $n \cdot \binom{2n}{2}$ possible types of pairs. If kn is greater than $n \cdot \binom{2n}{2}$, there must be two pairs of the same type, and we get the 4 squares we are looking for. This is the same as $k > \binom{2n}{2}$, so $\binom{2n}{2} + 1 = 2n^2 - n + 1$ is n-square.

To see that $2n^2 - n$ is not n-square, we consider a $2n \times (2n - 1)$ board. We are going to color it so that of all the possible $\binom{2n}{2}$ pairs of columns, there is exactly one row where they have the same color. If we can do this, then we can color in the same way $n - 1$ other boards of the same size, but rotating the colors, and then we paste them together. If these boards are $X_1, X_2, \ldots, X_n$, note that for each pair of columns, in each X_i there is exactly one row where they have the same color. Moreover, in each X_i this color is different, so these columns cannot give the 4 squares we want. This would give a coloring showing that $2n^2 - n$ is not n-square.

To do the coloring of X_1, first organize a tournament of $2n$ persons in $2n - 1$ rounds. In each round we split the persons into n pairs and they are going to play against each other. In the end we want each pair of persons to have played exactly once. If we can make this tournament, then the pairs of persons that played each other on the round j are going to represent the pairs of squares we paint the same color in the row j of the $2n \times (2n - 1)$ board. This gives the coloring we want (what color we give to what pair is irrelevant, we only have to give different colors to different pairs in the same row). This is because every pair of persons played each other in exactly one round, so their squares get the same color in exactly one row.

Thus we only have to show that such a tournament is possible. For this consider a regular polygon with $2n - 1$ sides, where each vertex represents one of the $2n$ persons. Note that there is person a that is not being represented. To do a round of the tournament, choose a vertex b of the polygon. The person in that vertex is going to play with a. The other $2n - 2$ vertices are going to be organized in pairs in the only way to make $n - 1$ parallel segments (those that have b in their orthogonal bisector). Those pairs play each other in that round. Doing this for every vertex in the polygon, we get the tournament we wanted.

Thus, the answer is $k = 2n^2 - n + 1$.

The solution might seem quite involved, since it use several transformations. However, making these tournaments via the polygon trick is very useful. Actually, it is used in this book in the solution of another problem, but we do not give the reference now so that readers that do not want any hints can enjoy the problem. □

---

[4]Or, if you are familiar with Jensen's inequality and the convexity of the function $f(x) = \binom{x}{2}$, we get $\sum_{i=1}^{n} \binom{n_i}{2} \geq n \left(\frac{\sum_{i=1}^{n} n_i}{n} \right) = n$. Here n_i is the number of squares of color i in the row.

Solution 2.24 Let a_i and b_i be the number of rows with color i and the number of columns with color i, respectively. Note that there are at most $a_i b_i$ squares of color i. Then we have that $N \le a_i b_i \le (\frac{a_i + b_i}{2})^2$. But since $N > n^2$, we have that $a_i + b_i > 2n$. Summing over i, we obtain $\sum_{i=1}^{N} a_i + \sum_{i=1}^{N} b_i > 2nN$. Thus, by the pigeonhole principle, either $\sum_{i=1}^{N} a_i > nN$ or $\sum_{i=1}^{N} b_i > nN$. Suppose without loss of generality that the first case holds. Note that $\sum_{i=1}^{N} a_i$ is the number of pair (C, i) where C is a column that has a square of color i. There are N possible columns and more than nN pairs. Thus there must be a column in more than n pairs, and that is what we wanted.   □

9.3   Solutions for Chap. 3

Solution 3.1 Note that if any of m or n is 1 we cannot construct the path. We are going to use the chessboard coloring. Two consecutive squares have different color. In a closed path there must be the same number of black and white squares. Thus if there is a closed path that uses all the squares once, the total number of white squares and the total number of black squares must be equal. Thus the number of squares is even, as we wanted.

Here is a path for a 6×7 board. This construction can clearly be extended if one of the side sizes is even and both are at least 2.

                                                                                     □

Solution 3.2 In the beginning the number of red and black tokens is different since 2009 is odd. Then A can remove some number of tokens of the color with more tokens so that the numbers of tokens of different colors become equal. If B plays, he cannot leave the same number of tokens of both colors, so he cannot win. Also, he permits A to repeat the process while reducing the number of tokens. Thus A wins after at most 2009 moves.   □

Solution 3.3 We will see that the losing positions are the odd numbers. We know that every divisor of an odd number is odd. So, if we have an odd number left, for the next step we leave an even number. If we have an even number left, we can remove exactly one coin. In this way we leave an odd number and we have not lost the game. Thus B has a winning strategy. □

Solution 3.4 Consider the following coloring:

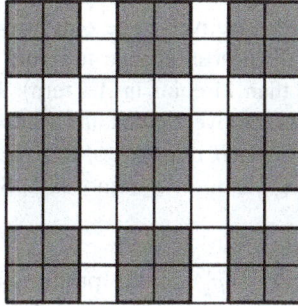

Note that when we do any change, we are changing an even number of lamps in the colored squares, so the number of lamps turned on in those squares remains even. Thus, if only one lamp remains on, it must be in a white square. To see that these can be changed it is sufficient to notice that they are at distance 2 from the edge. If X is the square at distance 2 from the edge and S, T are the squares separating it from the edge, we have the following situation.

We can use T and change the state of S, T, X and then use S and change the state of S, T. With this only X is turned on. When we used this pair of movements we have not affected any other lamp. As we wanted, X is the only lamp turned on in the whole board. □

Solution 3.5 Focus on a 5×1 piece. If we extend it by one of its sides to a 6×1 piece we have 3 colored squares there. If the 5×1 piece does not have 3 colored squares, then the coloring in the 6×1 piece must be alternating. This is impossible due to the second condition.

We can divide the board into a 1000×1001 board, a 1×1000 board and a 1×1 single square. The first two can be divided into parts of size 5×1 and 1×5 respectively, so the number of colored squares is at least $\frac{3(1001^2-1)}{5}$. To show that a coloring with this number is possible, we can do the following.

We start to color from left to right the first row using the figure above. Then we do the same thing in the next row but with the coloring shifted one square to the left, and so on. With this every piece of 1×5 has 3 colored squares. If we divide the board as indicated, the square that is left alone is not colored, so it causes no problems. Thus the number we wanted is $\frac{3(1001^2-1)}{5}$. $\qquad\square$

Solution 3.6 We will see that the set of losing positions is the set of numbers congruent to 0, 1, 4, 8 modulo 11. If A is in a losing position and removes 5 or 6 coins, B can remove 6 or 5 coins, respectively. The only way that this is impossible for B is if there are less coins than those needed to return A to the losing position it was in (there should be less than 11 coins in A's turn). The only case is if there are exactly 8 coins left, so B can remove 2 coins and win. If A removes 2 coins, there are several cases. If A was in 1, B removes 6. If A was in 0, B removes 5. If A was in 4, B removes 2. If A was in 8, B removes 5. Since 100 is congruent to 1 modulo 11, A loses. $\qquad\square$

Solution 3.7 Let $N = p_1^{a_1} p_2^{a_2} \cdots p_k^{a_k}$ be the prime factorization of N. In an arbitrary move the player writes down a divisor of N, which we can represent as a sequence $(b_1, b_2, \ldots, b_k)$, where $b_i \leq a_i$. Thus, a sequence $(b_1, b_2, \ldots, b_k)$ can be followed by $(c_1, c_2, \ldots, c_k)$ if either $0 \leq c_i \leq b_i$ for all i, or $b_i \leq c_i \leq a_i$ for all i. If at least one of the a_i is odd, then B has a winning strategy. Suppose that a_1 is odd. Split the sequences into pairs of the form $(b_1, b_2, \ldots, b_k)$, and $(a_1 - b_1, b_2, b_3, \ldots, b_k)$. After A plays any sequence, B plays the other in the pair. This is a winning strategy for B. If all a_i are even, we can use a similar argument to split the sequences different from $(a_1, a_2, \ldots, a_k)$ into pairs. If $(b_1, b_2, \ldots, b_k) \neq (a_1, a_2, \ldots, a_k)$, let j be the smallest index such that $b_j < a_j$. Then the pair of $(b_1, b_2, \ldots, b_k)$ is $(b_1, b_2, \ldots, b_{j-1}, a_j - b_j - 1, b_{j+1}, \ldots, b_k)$ (note that $b_j \neq a_j - b_j - 1$ since a_j is even). In the first turn A plays the sequence $(a_1, a_2, \ldots, a_k)$. After a move from B, A only needs to respond with the pair of the last sequence played. Thus the numbers for which A has a winning strategy are the perfect squares. $\qquad\square$

Solution 3.8 Count the pairs (a, b) of numbers $a \leq b$ that are not in order with respect to the one we want to reach. A horizontal movement of the empty space does not change the number of pairs. A vertical movement changes an odd number. However, since we want the empty space to end in its original position, the number of times it goes up must be equal to the number of times it goes down. Thus the number of vertical movements must be even. This means that the parity of the number of pairs must be preserved. Since it was 1 in the beginning, it cannot be 0 in the end. $\square$

Solution 3.9 We show that the losing positions modulo 13 are 0, 7, 5 without being able to remove 5, and 3 without being able to remove 3. Let us show that if we are in any of these positions the other player can keep us there. If we are in 0 and we remove some amount a different from 3, the other player can remove $6 - a$ and we are back in 7. If we remove 3, the other player can remove 5 and we are in 5 without being able to remove 5. If there are 7 left and we remove an amount a different from 1, the other player can remove $7 - a$ and we are back in 0. If we remove 1 the other player can remove 3 and there are 3 left without being able to remove 3. If we have 5 left without being able to remove 5, when we take out a, the other player can take out $5 - a$ and we are back in 0. If there are 3 left without being able to remove 3, then if we take 1 or 2, the other player can take 2 or 1 respectively. If we remove 4 or 5, the other can remove 5 or 4 and leave us in 7. Since we cannot get to 0 modulo 13, we cannot win. Note that if A begins by removing 4 coins, there are 7 modulo 13 coins left. Thus A has a winning strategy.   □

Solution 3.10 Note that 997 is prime. By removing the card $(1, 997)$, a player is going to lose if and only if he removes a card that does not have a multiple of 997. Note that there are exactly two multiples of 997 between 1 and 2003. There are 2002 cards that contain 997 and 2002 cards that contain 1994, but we are counting twice the card $(997, 1994)$. Thus, there are 4003 cards with a multiple of 997. Since this number is odd, every time B removes one of these cards A can remove another. Thus A has a winning strategy.

   Note that the only thing needed to do to obtain this strategy is finding a prime p with r multiples between 1 and 2003, where r is 2 or 3 modulo 4. Removing the card $(1, p)$ wins the games.   □

Solution 3.11 Consider the following coloring of the board:

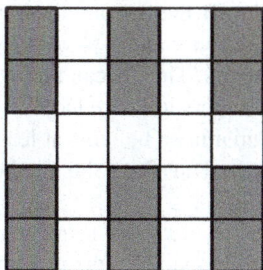

Note that by touching any lamp, we only change an even number of lamps in the colored squares. Thus, if a single lamp is turned on, it cannot be in any of these squares. By rotating that coloring 90 degrees, we have eliminated all possibilities except the following squares:

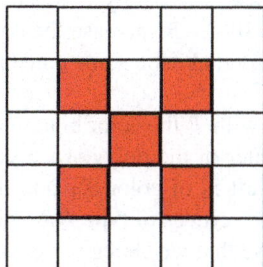

To leave exactly one of these lamps on here, we show the squares we have to touch.

In this problem, finding the lamps we need to touch to leave exactly one square turned on is quite hard. The reason for this difficulty is that the only way to get them is by trial and error.   □

Solution 3.12 First we design a strategy for B. In his first turn he removes the odd numbers, that way the minimum difference between two numbers left is at least 2. In his second turn there are only numbers of the type $4k$ and $4k + 2$. As B has to remove half of the numbers, he can guarantee that he can remove all the ones of one of these two types. Thus the minimum difference between two numbers is now at least 4. B can keep on doing this and doubling the minimum difference at each turn.

The strategy for A is similar. If the two numbers with the biggest difference at some moment are x and y, between x and $\frac{x+y}{2}$ or between $\frac{x+y}{2}$ and y we have half or less than half of all the numbers. Thus A can remove all of these. The maximum difference between any two numbers left is reduced at least by half.

Using this strategy B can guarantee he wins at least 32 dollars and A can guarantee he loses at most 32 dollars. Thus 32 is the number we wanted.   □

Solution 3.13 Denote by $x_1, x_2, \ldots, x_N$ the sequence of numbers in question. We call the "completed length" of the sequence the maximum number k such that $x_1 \leq x_2 \leq \cdots \leq x_k$. B wants to change the numbers so that the completed length is N. Suppose that k is the completed length, $k < N$, and A gives a certain number r. If $x_k \leq r$, then B can replace x_{k+1} by r and the completed length increases by 1. If not, then there must be a $0 < t \leq k$ such that $x_{t-1} \leq r < x_t$ (consider $x_0 = 0$ for this purpose). If B changes x_t by r, then the completed length remains the same, but the sum of all the numbers decreases at least by 1. Since the sum of all the numbers is non-negative, A must eventually let B increase the completed length. If B follows this strategy, we will eventually make the game come to an end.   □

Solution 3.14 We abbreviate with B the color blue, R the color red and Y the color yellow. If we stand at the center of the polygon we say that a vertex is of the type $X_1 X_2$ if the segment to its left is of color X_1 and the segment to its right is of color X_2. We assign $+1$ to the vertices of type BR, RY, YB and -1 to the vertices of type RB, BY, YR. Suppose that we change a segment from yellow to red. Then its two neighbor segments had to be blue. Thus we are changing vertices of types BY, YB to BR, RB. Note that the total sum does not change. Since the sums of the two colorings we are given are different, we cannot get from one to the other.   □

Solution 3.15 We are going to count the number of pairs (a, C) where C is a 2×2 square and a is a white unit square of C. If what we want is not satisfied, each square C is in an even number of pairs, so the total number of pairs should be even. However, every white square in the sides is in 2 pairs and every white square in the interior is in 4 pairs. Since there are 3 white squares in the corners and each is in exactly one pair, the total number of pairs must be odd. Thus we get a contradiction, as we wanted. □

Solution 3.16 Consider the set X of squares which were originally white and the set Y of squares that were originally black. Assign a 0 to each white square, a 1 to each black square and a 2 to each green square. Note that each time we are applying the operation we are using one square of X and one of Y. Thus the sum modulo 3 of the numbers assigned to the squares in X grows by 1 after each operation, and so does the sum of the squares in Y. We also want each square in X to be 1 at the end, so if x is the number of squares of X, the number of operations we need is congruent to x modulo 3. Since we want each square of Y to end up with 0, the corresponding numbers must go down by 1 modulo 3 each. If there were y squares in Y originally, the number of moves has to be congruent to $-y$ modulo 3. That is, $x \equiv -y \pmod{3}$, which is the same as $x + y \equiv 0 \pmod{3}$. But $x + y$ is the total number of squares, so we need one of the numbers a, b to be a multiple of 3.

If any of a, b is a multiple of 3 we can divide the board into 3×1 strips. In each of these we can swap the colors in the following way:

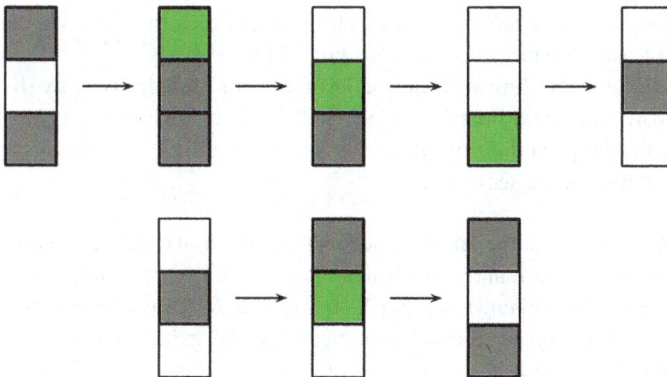

We conclude that the change can be done if and only if one of a and b is a multiple of 3. □

Solution 3.17 First we will see that A has the winning strategy. Note that the odd factors of the initial pile of coins are always going to be preserved. Thus we can think that the game starts with one pile having a power of 2 greater than or equal to 4.

Let T be the number of piles whose number of coins is a multiple of 4. If we split a pile that is multiple of 8, T increases by 1. If we split a pile with 4 coins, then

T decreases by 1. Thus, unless we remove the odd piles or split a pile with 2 coins, the parity of T changes.

A plays to leave B with an even T and no odd piles. If in his turn B changes the parity of T, then T has to be odd and A can change it again. If in its turn B splits a pile with 2 coins, then A removes the odd piles. B can never win if A plays this way, and eventually the game ends. Thus A has a winning strategy.

Once we know this, finding the conditions under which A wins is easier. If T is odd and there are no odd piles, A wins. If T is even but there are odd piles, A also wins by removing the odd piles. If T is even and there are no odd piles, then A loses (B plays as before). If T is odd and there are odd piles, then A cannot change the parity of T or remove the odd piles, since that would leave B in one of the previous situations where he would win. Thus A can only split piles with even number of coins but not multiples of 4. The same thing happens for B. For A to win in this position we need an odd number of even piles that are not multiple of 4.

Thus, the conditions under which A has a winning strategy are

- T is odd and there are no odd piles;
- T is even and there are odd piles;
- T is odd, there are odd piles, and the number of even piles that are not multiples of 4 is odd.   □

Solution 3.18 We will color the rows black and white alternatingly starting with black. There are 2000 more black squares than white squares. The shapes of the first two types always use the same number of black and white squares. The shapes of the third type always use 3 squares of one color and 2 of the other one. This means that we need at least 2000 pieces of the third kind. Thus, $s \leq \frac{1993 \cdot 2000 - 5 \cdot 2000}{4} = 994000$.

We can produce a tiling with only 2000 pieces of the third type by dividing the original board into 1000 boards of size 19993×2. In each of these we place two tiles of the third type at the extremes with their tips on different columns and we fill the rest with tiles of the second type.   □

Solution 3.19 Note that the empty space changes color at every turn, so it is always black when it is A's turn and it is always white when it is B's turn. Thus A always has to move a white token into a black square and B always has to move a black token into a white square. Hence, it is clear that the game cannot go on after 12 moves from each player.

Now we show that B has a winning strategy. Regardless of which token A removes at the beginning, the rest of the board can be tiled with 2×1 domino tiles. This is because the first token A removed was black. Since A moves for the first time, he is removing one token from a domino tile, so B can move the other token in that domino along it. Thus in the next turn A has to remove again a token from a domino tile, and B can move along it. If B follows this strategy, he can always move if A could do it before. Thus, B has a winning strategy.   □

Solution 3.20 Given an orthogonal polygon with integer sides we can give it a chessboard coloring. Let b and w the number of black and white squares, respectively.

Let B and W be the length of the boundary that ends up painted black and the length of the boundary that ends up painted white, respectively. We will show that $4(b - w) = B - W$. For this consider any vertical strip of the polygon.

Note that if in the strip there is the same number of black and white squares, then the two horizontal segments of the boundary that bound this strip are of different colors. If there are more squares of one color (there can only be exactly one more) then the two horizontal segments we mentioned have that color. If a vertical strip intersected several disjoint sections of the polygon, this argument can be done separately for each of them. If B_H and W_H are the lengths of the horizontal black and white segments of the boundary (respectively), we can add up by the vertical strips and get $2(b - w) = B_H - W_H$. Doing the same thing for horizontal strips and adding up we get $4(b - w) = B - W$.

Once we have this, consider a polygon with only odd sides. When we give it a chessboard coloring, all the corners have the same color. Thus, all the sides have one more square of that color. Therefore, $B - W \neq 0$, which means that $b - w \neq 0$. This implies that the polygon cannot be tiled with domino tiles.   □

Solution 3.21 In his first turn B places an S far away from the first letter of A and the edges (at least 6 squares of distance from each). If he cannot win in his second turn, he places an S three squares away from his previous S, but without giving A the chance to make an SOS. This is always possible, because B has two places where he can write the second S, and the second letter of A can only be near one of them. Thus we have two squares with S separated by two empty spaces. If any player writes anything in those two squares the other can win, so the game cannot end in a draw.

After this, the strategy for B is as follows. If he can make an SOS, he does and wins. If not, he searches for a square that has both its neighbors occupied or both its neighbors free, and writes an O there. If he could not find that square it means that every empty square has exactly one empty square as a neighbor. That is, there is an even number of empty squares. This can never happen in B's turn, since they must have played an odd number of times before him.

Note that by playing like this A cannot win. This is because since B could not make an SOS in his turn, A would need to use the letter B just wrote to win. However, the way B chooses to write his O makes that impossible. Thus B has a winning strategy.   □

Solution 3.22 First notice that every tile has a hole, which we call the center of the tile.

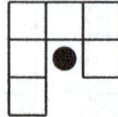

Note that if a tile B covers the center of a tile A, then A covers the center of B. Hence, the tiles come in pairs with one of the following two shapes:

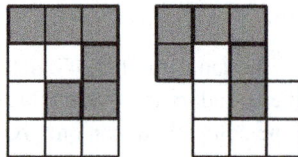

Thus mn has to be a multiple of 12. We will show that at least one of the numbers m or n has to be a multiple of 4. If this does not happen, since their product is a multiple of 4, then both have to be even. If we number the rows and columns and add 1 to each square if its column is divisible by 4 and then add 1 to each square if its row is divisible by 4, we end up with a coloring of the following type:

			1				1		
			1				1		
1	1	1	2	1	1	1	2	1	1
			1				1		
			1				1		
			1				1		
1	1	1	2	1	1	1	2	1	1
			1				1		
			1				1		
			1				1		

Note that the total sum of the written numbers is even (since we added an even number of 1s in each step). It is easy to see that, regardless of how we place any of the two figures we have, the numbers they cover have an odd sum. Thus there should be an even number of tiles, which means that mn is a multiple of 24. Thus, one of the sides must be a multiple of 4.

We can also note that if any of m, n is 1, 2 or 5, then we cannot tile the board. If any of them is 3, then the other is a multiple of 4 and we can tile it with 3×4 tiles. The same thing happens if any of them is 4, so we may assume that they are both at least 6.

Note that $6 = 3 + 3, 7 = 4 + 3$ and $8 = 4 + 4$. Any integer greater than or equal to 6 can be written in the form $s + 3r$, where s is one of 6, 7, 8 and r is a non-negative integer. Thus, any integer greater than or equal to 6 can be written as $3a + 4b$ with a and b non-negative integers.[5]

If any of n, m is multiple of 3 and the other is multiple of 4, we can clearly tile the board with 3×4 tiles. If one is multiple of 12 (suppose it is m) and the other is at least 6, it can be written as $3a + 4b$. Thus the board can be split into a strips of size $3 \times m$ and b strips of size $4 \times m$. Since each strip can be tiled using 3×4 tiles, we can tile the whole board.

Thus the pairs (m, n) that work are those where mn is a multiple of 12, at least one of them is divisible by 4 and neither of them is 1, 2 or 5.   □

Solution 3.23 Let us color the 5 rows black and white alternatingly. Note that the proportion of black to white is $3 : 2$, so all tiles must have 3 black squares and 2 white ones (if not, then the total proportion cannot be obtained). Also, every tile is like a 2×2 square with a tip. We have shown that all tips must be black. Thus the white strips must only be covered with the 2×2 part of the tiles. This means that n must be even.

Let a_k be the number of ways to tile a $5 \times 2k$ board with those tiles and b_k be the number of ways to do it without being able to split the board into two smaller boards of sizes $5 \times 2k_1$ and $5 \times 2k_2$, each tiled as the problem asks.

Note that $b_1 = a_1 = 2$ and that in a board we can place in its right edge the following pieces:

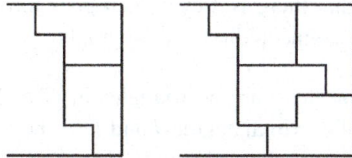

To each of these we can add the following ones:

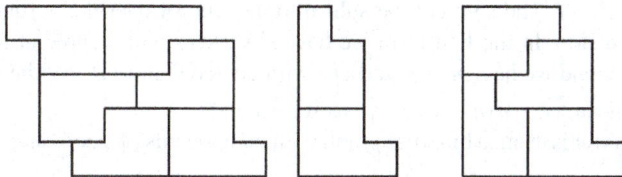

---

[5]The interested reader may want to prove that given two positive integers x, y that are relatively prime, the largest integer that cannot be written as $ax + by$ with a, b non-negative integers is $ab - a - b$.

We can use the first piece to enlarge our tiling until this is no longer possible. Then, we can finish with one of the two last pieces (if they do not fit, changing the piece used in the beginning can fix it). With this method we can cover any $5 \times 2k$ board without being able to draw a vertical line that cuts no tiles. Since we can flip this cover horizontally, it means that it can be done in at least two ways. Suppose that for all $r < k$ we have that $a_r \geq 2 \cdot 3^{r-1}$. If we count the number of ways to cover a $5 \times 2k$ board by the place where we can draw the first vertical line (from left to right) that cuts no tiles, we get that $a_k = b_1 a_{k-1} + b_2 a_{k-2} + \cdots + b_{k-1} a_1 + b_k$. Thus, using our induction hypothesis and Exercise 1.2.5 to simplify, we obtain $a_k \geq 2(2 \cdot 3^{k-2}) + 2(2 \cdot 3^{k-3}) + \cdots + 2(2) + 2 = 2[2 \cdot (1 + 3 + 3^2 + \cdots + 3^{k-2}) + 1] = 2[3^{k-1} - 1 + 1] = 2 \cdot 3^{k-1}$. $\qquad\square$

9.4   Solutions for Chap. 4

Solution 4.1 Let $v_1, v_2, \ldots, v_k$ be the longest path in the graph. v_1 is adjacent to at least three vertices, so at least to two vertices different from v_2. These two vertices must be in the path, or it would contradict its maximality. Then v_1 is adjacent to v_2, v_i, v_j. By the pigeonhole principle, two of the numbers $2, i, j$ have the same parity. The section of the path between their corresponding vertices and v_1 make the cycle we wanted. $\qquad\square$

Solution 4.2 Let N be the connected component with the smallest number of vertices. If G is not connected, then it must have at least two connected components. Thus N has at most $\frac{n}{2}$ vertices. Since vertices in N can only be adjacent to vertices in N, they must have degree smaller than or equal to $\frac{n}{2} - 1$, which contradicts the hypothesis.

In a graph with this condition we can actually get a much stronger statement. Namely, one can prove that there is a cycle that goes through every vertex once,[6] which in turn implies the connectedness of the graph. $\qquad\square$

Solution 4.3 Suppose that there are no triangles in G and let e be the number of edges. Let u be a vertex of maximal degree d and $\Gamma(u)$ be the set of vertices adjacent to u. As there are no triangles, the vertices of $\Gamma(u)$ form an independent set (there are no edges among them), so their degree is at most $n - d$. The other $n - d$ vertices outside $\Gamma(u)$ (these include u) have degree at most d. Thus, by Example 4.1.1, $2e = \sum_{k=1}^{n} d(v_k)$. The sum can be split into the sum of over the vertices in $\Gamma(u)$ and those outside. In the first term we have d vertices with degree at most $n - d$ and in the second we have $n - d$ vertices with degree at most d. By the arithmetic-geometric mean,[7] $e \leq d(n - d) \leq \frac{n^2}{4}$, as we wanted.

Note that if n is even, a bipartite graph with components of size $\frac{n}{2}$ and all possible edges has no triangles and exactly $\frac{n^2}{4}$ edges. $\qquad\square$

---

[6]This is called a **Hamiltonian cycle**.

[7]This inequality states that for any two non-negative real numbers a, b we have $\sqrt{ab} \leq \frac{a+b}{2}$.

Solution 4.4 Consider a graph with 28 vertices and 14 edges representing the games. To each vertex we can assign a player, depending on whether we was in that game or not. Since every player has played at least once, to 20 of those vertices we can assign different players. There are 8 vertices left to mark, but those use at most 8 edges. Thus, the other 6 edges are full and use 12 different players. □

Solution 4.5 Consider a graph with one vertex per person and an edge if the two persons are seated with exactly one person between them. What we get is the union of two disjoint cycles of length $3n$ (the persons seated in an even position and the ones seated in an odd position). We color the vertices with 3 colors according to nationality. Note that, if there are two adjacent vertices of the same color, then the person between them is standing up, and if not the person between them remains seated. Thus we want to maximize the number of adjacent vertices of the same color. Consider one of the cycles. As there are only $2n$ vertices of each color, it cannot be monochromatic. Consider a vertex v_0 in that cycle. Start in v_0 and go around the cycle until you reach a vertex of another color. You can do this in two ways, one for each direction in the cycle. Note that when you find a vertex of different color for the first time you use a different edge depending on which sense you chose to go along the cycle.

Thus in that cycle we have at least 2 seated persons. If we do the same for the other cycle we get that at most $6n - 4$ persons are standing up. To see that this number can be reached, paint in one cycle $2n$ consecutive vertices red, in the other $2n$ consecutive vertices green, and the rest of the vertices blue. Then seating the people according to nationality we obtain an arrangement where only 4 persons stay seated. □

Solution 4.6 First Solution Consider a graph with one vertex per person, a red edge if two persons know each other, and a blue edge if they do not. Let A be the largest set of vertices with only red edges and x, y be two vertices outside A. We want to show that the edge xy is blue.

First note that there must be vertices in A that are connected to x with a blue edge, or we could add x to A, contradicting its maximality. The same goes for y. Let us assume that the edge xy is red and look for a contradiction. Consider all the vertices in A that are connected either to x or y with a blue edge. If there is only one such vertex, we can remove it and add x and y to A, contradicting it maximality. Thus there are two vertices z, w in A such that the edges (w, x) and (z, y) are blue. This means that in the set $\{x, y, z, w\}$ we cannot find 3 vertices that satisfy the condition of the problem. This is the contradiction we wanted. □

Second Solution We solve the problem by induction on n. If $n \leq 4$ the assertion is true. Suppose it is true for a certain k and let us prove it for $k + 1$. For this, construct a graph as in the previous solution. Let v_0 be an arbitrary vertex. Since the other k vertices satisfy the condition of the problem, we can split them into two sets, A and B, so that in A there are only red edges and in B there are only blue edges. Choose that splitting so that the number T of blue edges from v_0 to A plus the number of red edges from v_0 to B is minimal.

Suppose that we cannot add v_0 to A. This means there is a vertex a in A such that $v_0 a$ is blue. The same way there must be a vertex b in B so that bv_0 is red. Suppose without loss of generality that ab is red. Let x be any other vertex of A. This means that xa is red. If bx was blue, then the vertices $\{v_0, a, b, x\}$ would not satisfy the problem's condition. Thus bx is red. Since this happens for every vertex in A, we can place b in A. With this we are reducing the number of red edges from v_0 to B, contradicting the minimality of T. Thus v_0 can be added to one of the parts and we are done.   □

Solution 4.7 We show by induction on the number of vertices of the graph that it is always possible to turn all the vertices black. Let n be the number of vertices of G. If n is 1, we only change that vertex. Suppose that we can do it for $n - 1$ and we want to prove that we can do it for n. Given an arbitrary vertex v, consider the graph induced by the other $n - 1$ vertices. By hypothesis, there is a series of movements that makes all the vertices in that graph black. If we apply those movements to G there are two possibilities. Either v becomes black or it does not. If v is black then we are done. If not, we can do this for every vertex in G. Thus, if it is not possible to do the change the problem asks to, for every vertex p there is a way to change the color of all vertices except p.

If n even, we can apply this new operation for every vertex. Each vertex changed color $n - 1$ times, so we end up with a completely black graph. If n is odd, then by Exercise 4.1.2 there must be a vertex v_0 of even degree. Let B be the set formed by v_0 and all its neighbors. If we perform the new operation for every vertex in B, every vertex outside has changed color and the vertices in B remain white. Then we can use the original movement on v_0 and we are done.   □

Solution 4.8 Consider a graph G with one vertex per person and an edge if two persons are friends. Note that since the group of friends is 3-indivisible, it is also 2-indivisible. This means that G is not bipartite. By Theorem 4.3.3, G has an odd cycle. Let B be the shortest odd cycle and C the rest of the vertices (if any). Note that B is not bipartite. We only have to prove that C is not empty and contains an edge. By the minimality of the cycle, there are no edges in B except for those of the cycle. Also, if a vertex c in C is adjacent to two vertices of B, they divide the cycle into an odd arc and an even arc. Since B is minimal, if we replace the vertices inside the even arc by c, we are not reducing its length. Thus, the two vertices c is adjacent to are at distance 2 in the cycle. If a vertex c was adjacent to at least three vertices of B, the previous statement would mean that B is a triangle and c is adjacent to all the three vertices. However, the conditions of the problem make this impossible. Thus every vertex in C is adjacent to at most two vertices of B.

B can be split into three independent sets X_1, X_2, X_3. Every vertex of C can then be placed in one of those sets so that it is not adjacent to a vertex of B in its set. Since the group of friends was 3-indivisible, there must be an edge in one of the X_i. By the construction of the sets, this edge must be in C, as we wanted to prove.   □

Solution 4.9 Consider a graph where every vertex represents a person and an edge represents a pair of persons that know each other.

Suppose that there are 4 vertices x_1, x_2, x_3, x_4 that form a 4-clique. Then, as any two 3-cliques have at least a vertex in common, any triangle must have two vertices of x_1, x_2, x_3, x_4. If there are no other triangles, then we can remove two of them and we are done. If there is another triangle, we can suppose without loss of generality that it has vertices x_1, x_2, x_5. We will show that if we remove x_1 and x_2 then there are no more triangles left. If not, there must be a triangle that only uses x_3, x_4 out of x_1, x_2, x_3, x_4. Since that triangle must have a vertex of x_1, x_2, x_5, it must be x_3, x_4, x_5. Thus x_1, x_2, x_3, x_4, x_5 is a 5-clique, which contradicts the hypothesis.

Thus there can be no 4-cliques. If the statement of the problem is false, then for any two vertices of the graph there must be a triangle that does not use them. Consider a triangle x_1, x_2, y. There must be a triangle that does not use x_1 or x_2, say x_3, x_4, y. We know there is a triangle that does not use y or x_2, so it must have x_1 and one of x_3, x_4. Using this we can suppose, without loss of generality, that x_1 and x_3 are adjacent. Since x_1, y, x_3, x_4 do not form a 4-clique, x_1 and x_4 cannot be adjacent. In the same way we conclude that one of x_1, x_2 is adjacent to x_4, so it must be x_2. Also, x_2 and x_3 do not form an edge, or we would have a 4-clique. The graph induced by x_1, x_2, y, x_3, x_4 must look like this:

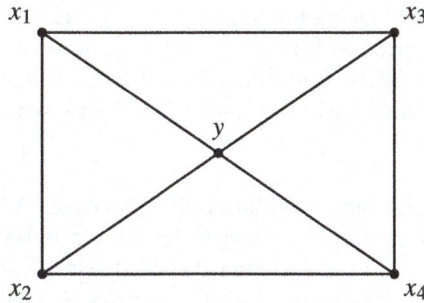

But, if we consider a triangle that does not have x_1 or y, it must have x_2 in order to intersect triangle x_1, x_2, y and it must have x_3 in order to intersect triangle x_1, y, x_3. Since x_2 and x_3 are not adjacent, this is impossible. $\square$

Solution 4.10 Consider a graph with one vertex per tourist and an edge between two vertices if the corresponding tourists know each other. The first condition says there are no triangles. the second says that the graph is not bipartite. Since the graph is not bipartite, by Theorem 4.3.3 there is at least one odd cycle. Let $(v_0, v_1, v_2, \ldots, v_k = v_0)$ be the shortest odd cycle. Since there are no triangles, $k \geq 5$. Suppose that a vertex outside of the cycle is adjacent to two vertices of the cycle, v_i and v_j. We know that v_i and v_j divide the cycle into two parts, one of even length and one of odd length. If we replace the even part by (v_i, v, v_j) we obtain a shorter odd cycle, unless v_i and v_j are at distance 2 in the original cycle. Thus v can only be adjacent to two vertices of the cycle. Also, the cycle cannot have diagonals because of its minimality, so every vertex of the graph (including the ones in the cycle) is adjacent to at most two vertices of $v_1, v_2, \ldots, v_k$.

Consider the pairs (u, v_j) where u is a vertex of the graph and v_j is a vertex of the cycle such that they form an edge. Every vertex u is in at most two pairs, so there are at most $2n$ pairs. However, since there are k vertices in the cycle, there must be one in at most $\frac{2n}{k}$ pairs. Thus, at least one of the vertices $v_1, v_2, \ldots, v_k$ has degree smaller than or equal to $\frac{2n}{k} \leq \frac{2n}{5}$. $\qquad\square$

Solution 4.11 To count the minimum number of cyclic sets, consider the teams in a line and a tournament where every team beats the teams on its right and loses to the teams on its left. This shows that there may be zero cyclic sets.

To find the maximum number of cyclic sets, note that in every non-cyclic set there is exactly one team that beat the other two. Number the teams from 1 to $2n + 1$ and denote by r_k the number of teams that were beaten by team k. Then the number of non-cyclic sets is $\binom{r_1}{2} + \binom{r_2}{2} + \cdots + \binom{r_{2n+1}}{2}$. Using Exercise 1.1.19 several times we obtain

$$\binom{r_1}{2} + \binom{r_2}{2} + \cdots + \binom{r_{2n+1}}{2} \geq (2n+1)\binom{\frac{r_1+r_2+\cdots+r_{2n+1}}{2n+1}}{2} = (2n+1)\binom{n}{2}.$$

Thus the number we are looking for is at most $\binom{2n+1}{3} - (2n+1)\binom{n}{2} = \frac{n(n+1)(2n+1)}{6}$.

To obtain a tournament with this number of cyclic sets it is sufficient to find one where $r_k = n$ for all k. To do this, place the $2n + 1$ teams on a circle and consider each team beating the next n teams in clockwise sense, and losing with the other n. $\qquad\square$

Solution 4.12 Consider a bipartite graph with components X and Y. X is going to have a vertex for each S_i and Y is going to have a vertex for each element of the union of the S_i. The pair $\{S_i, s\}$ is going to be an edge if s is an element of S_i. Note that given a subset $\mathcal{S} \subset X$, $\Gamma(\mathcal{S})$ is the set of vertices that have an edge to at least one element of $\mathcal{S}$. In other words, it consists of the vertices representing an element that belongs to some set of $\mathcal{S}$. Hence, $\Gamma(\mathcal{S})$ represents the elements belonging to the union of the sets in $\mathcal{S}$. Thus, the problem is equivalent to Hall's marriage theorem. $\square$

Solution 4.13 Let d be the common degree and X and Y be the two components of the graph. Note that since every edge has exactly one vertex in X, the total number of edges is $d|X|$. In the same way, the number of edges is $d|Y|$. Thus, X and Y have the same number of vertices. Also, given $S \subset X$ with s elements, from S we have ds edges. Thus we cannot have less than s elements in $\Gamma(S)$. Since we have the conditions for Theorem 4.4.2, we are done. $\qquad\square$

Solution 4.14 We will show that we can extend the board to an $n \times (k + 1)$ board with the same properties. For this, consider a graph with $2n$ vertices. n of them are going to represent the rows of the board and n the numbers from 1 to n. We place an edge between a row and a number if the number is not yet present in that row. Note that every row contains k numbers, so their degrees in the graph is $n - k$. However, every number appears exactly once in each column, so they are in k different rows.

Thus their degrees are also $n - k$ in the graph. Using Problem 4.13, the graph has a perfect matching. This matching tells us which numbers to write in the $(k + 1)$-th row so that the property is not violated. We can repeat this process until we get to an $n \times n$ board. □

Solution 4.15 Consider a graph G whose vertices represent the points and there is an edge between two vertices if the interior of one of the arcs they define has exactly n vertices. We want to show that when we color k vertices, two adjacent vertices have been colored. Since the degree of each vertex is 2, Exercise 4.1.4 shows that G is the union of disjoint cycles. Note that if we number the vertices from 1 to $2n - 1$, then 1 is adjacent to $n + 2$ and $n + 2$ is adjacent to $2n + 3$, which is vertex 4. Thus 1 and 4 are in the same cycle. If $2n - 1$ is not divisible by 3, then G consists of only one cycle, so $k = n$ is clearly the desired number. If $2n - 1$ is divisible by 3, then the graph is formed by three disjoint cycles of length $\frac{2n-1}{3}$. Thus we can color at most $\frac{\frac{2n-1}{3}-1}{2} = \frac{n-2}{3}$ vertices of one cycle without getting consecutive vertices colored. Thus $k = 3 \cdot ((n-2)/3) + 1 = n - 1$ is the number we want in this case. □

Solution 4.16 Consider two adjacent vertices v_i and v_j with degrees $d(v_i)$ and $d(v_j)$. From v_i there are another $d(v_i) - 1$ edges and from v_j there are another $d(v_j) - 1$ edges. Since there are only $n - 2$ other vertices, v_i and v_j must be adjacent to at least $[d(v_i) - 1] + [d(v_j) - 1] - (n - 2) = d(v_i) + d(v_j) - n$ vertices in common. Then there are at least $d(v_i) + d(v_j) - n$ triangles that use the edge (v_i, v_j). If we do this for every edge and we sum, we obtain the following number of triangles

$$\sum_{v_i, v_j \text{ adjacent}} [d(v_i) + d(v_j) - n].$$

Note that in every term we have a $-n$ and that every term $d(v_i)$ appears for every edge that has v_i as vertex (that is, $d(v_i)$ times). Thus, we can write the sum as

$$\left[\sum_{i=1}^{n} d(v_i)^2 \right] - nk.$$

We know that $\sum_{i=1}^{n} d(v_i)^2 \geq \frac{(d(v_1)+d(v_2)+\cdots+d(v_n))^2}{n}$. Using the fact that $d(v_1) + d(v_2) + \cdots + d(v_n) = 2k$, we are counting at least $\frac{4k^2}{n} - nk = \frac{k(4k-n^2)}{n}$ triangles. However, every triangle is being counted at most three times (once for each of its edges), so there are at least $\frac{k(4k-n^2)}{3n}$ triangles in the graph.

Note that this number is positive if $k > \frac{n^2}{4}$, which was the case in Problem 4.3. □

Solution 4.17 First note that $110 = 6(18) + 2$. We will show that if $6k + 2$ teams or more play in 6 rounds, then we can find $k + 1$ teams that have not played one another. If we consider a graph with one vertex per team and an edge between two teams that have played once, the problem reduces to the following: "If G is a graph with $6k + 2$ vertices that is the union of 6 perfect matchings, there exists an independent set of $k + 1$ vertices".

If $k = 1$ we can note that every team has played with 6 teams, so if we fix a team p_1, we can find another team that did not play a match against p_1. These two teams are the ones we are looking for. Now suppose the statement is true for all $r < k$ and we want to prove it for k.

Consider again an arbitrary team p_1. Then p_1 has played with 6 other teams. Let P_1 denote the set of teams that p_1 played with. Note that we have considered 7 teams, so they cannot have played all their games between them. Thus there is a team p_2, different from all the teams we have considered, that has played with a team of P_1. Let P_2 denote the set of all the new teams that p_2 has played with. If we can find a team different from the ones we have considered that has played with a team of P_1 or P_2, we call it p_3 and call P_3 all the new teams p_3 has played with.

Suppose that we can keep on generating this sequence of teams $p_1, p_2, p_3, \ldots$. Note that P_1 has 6 teams. P_2 has at most 5, P_3 has at most 5 and so on. So, when we generate p_t, we have considered at most $6t + 1$ teams. Thus we can obtain $k + 1$ teams $p_1, p_2, \ldots, p_{k+1}$ that have not played each other.

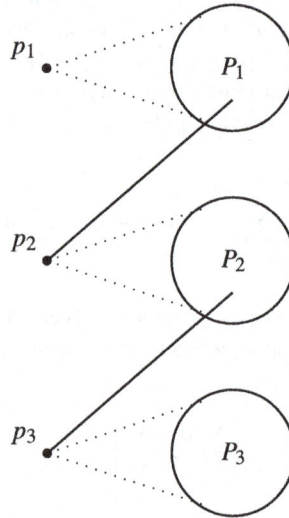

If we cannot keep on doing this list, it means that there is an r such that there is no possible choice for p_{r+1} in the rest of the graph. That means that all the teams we have considered up to P_r have played all their matches between themselves. Note that we have considered at most $6r + 1$ teams, but they must be an even number to have played all their games between themselves. Thus there are at most $6r$ of them. Then there are at least $6(k - r) + 2$ other teams that have played all their games between themselves. By the induction hypothesis, we can find $k - r + 1$ teams that have not played each other. If we add $p_1, p_2, \ldots, p_r$ to this set, we have the $k + 1$ teams we wanted. $\square$

Solution 4.18 Consider a graph G where every vertex represents a city and every edge a road between the corresponding cities. Denote by t the "radius" of the graph G. Let H be the subgraph of G with the same radius but with the minimum number of vertices possible. We will solve the problem in H. Given vertices x, y in H denote by $d(x, y)$ the minimum number of edges one needs to use to get from x to y. That is, $d(\cdot, \cdot)$ is the distance in the graph. By Exercise 4.2.8, there is a vertex v_t such that if we remove it the graph is still connected.

By the minimality of H, when we remove v_t the resulting graph cannot have radius t. As the graph is still connected it must have a smaller radius. That is, there is a vertex v_0 such that $d(v_0, w) \leq t - 1$ for all w different from v_0. Then $d(v_0, v_t) = t$. Let $(v_0, v_1, v_2, \ldots, v_t)$ be a path of length t from v_0 to v_t. This is why v_t was given this name. We have that v_i and v_j are adjacent if and only if i and j are consecutive.

If $t = 2$, then v_0, v_1, v_2 are the vertices we are looking for, so we may suppose that $t \geq 3$ (if $t = 1$ any vertex works). As the graph H has radius t, there is a vertex w such that $d(v_2, w) = t$. With this, we have that $w \neq v_i$ for all i. We also know that $d(w, v_0) \leq t - 1$. Consider a path R of minimal length that joins v_0 with w. Let u be any vertex of R. Suppose that $d(u, v_i) = 1$ for some $i \geq 2$. Let $p = d(v_0, u)$ and $q = d(u, w)$. Then,

$$i = d(v_0, v_i) \leq d(v_0, u) + d(u, v_i) = p + 1,$$
$$t = d(v_2, w) \leq d(v_2, v_i) + d(v_i, u) + d(u, w) = (i - 2) + 1 + q = i + q - 1.$$

If we add these two inequalities we have that $t \leq p + q = d(v_0, w) = t - 1$, which is a contradiction. Consequently, $u \neq v_1$, so the path from v_0 to w does not go through v_1 (or any v_i with $i \geq 1$).

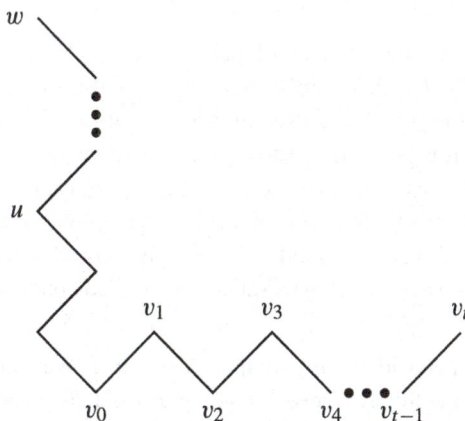

We know that v_1 is adjacent to v_0, so we can consider the vertex u_0 of R that is closest to w and adjacent to v_1 (if there is no other, $v_0 = u_0$). Then $d(u_0, v_1) = 1$. Moreover,

$$t = d(v_2, w) \leq d(v_2, v_1) + d(v_1, u_0) + d(u_0, w) = 1 + 1 + d(u_0, w),$$

whence $d(u_0, w) \geq t - 2$. Thus the path we are looking for is

$$w \cdots u_0 v_1 \cdots v_t.$$

$\square$

9.5   Solutions for Chap. 5

Solution 5.1 By the pigeonhole principle there are two elements from $\{1, 2, \ldots, n+1\}$ that are going to the same element. However, there cannot be any other pair that does this or the function would not be surjective. There are $\binom{n+1}{2}$ ways to choose this pair. Knowing this, we can treat the pair as if it was an element and we only have to find the number of bijective functions for two sets with n elements each. There are $n!$ of these, so the number we are looking for is $\binom{n+1}{2}n!$. $\square$

Solution 5.2 Suppose that we have a group of n persons. We are going to consider the number of ways to make a committee that has one president and one secretary who is different from the president. If the committee is going to have k persons in total, there are $\binom{n}{k}$ ways to choose the members, and then we have k options for the president. After that, we have $k - 1$ options for the secretary. However, we can choose first the president, and then the secretary, which can be done in $n(n-1)$ ways. Then the remaining $n - 2$ persons may or may not be in the committee. Thus there are $n(n-1)2^{n-2}$ possible committees. This gives $\sum_{k=2}^{n} k(k-1)\binom{n}{k} = n(n-1)2^{n-2}$. $\square$

Solution 5.3 Let T be the number of pairs (P, c) such that P is a clown and c is a color used by P. Then, since every color is used by at most 20 clowns, $T \leq 20 \cdot 12$. Since every clown uses at least 5 colors, $T \geq 5n$. It follows that $n \leq 48$. To see that it is possible to satisfy the conditions with 48 clowns, consider the following 5-color sets $(c_1, c_2, c_3, c_4, c_5)$, $(c_1, c_2, c_3, c_4, c_6)$, $(c_1, c_2, c_3, c_4, c_7)$, $(c_1, c_2, c_3, c_4, c_8)$. Consider the sets obtained by taking one of these and adding to the indices modulo 12 a certain number i. This gives us 48 sets that can be used to dress the 48 clowns. These 48 clowns satisfy the needed conditions. $\square$

Solution 5.4 We can consider that both lists start with 1. We add another row to the board. The sums of the first two rows are going to yield 2003 (or 2004) consecutive numbers if and only if in the third row we can place again the number from 1 to 2003 (or 2004) so that the sum by columns is constant. With 2004, let t be the column sum. $2004t$ is the total sum, which corresponds to 3 times the sum of all numbers from 1 to 2004. Then $2004t = 3(\frac{2004 \cdot 2005}{2})$, so $t = \frac{3 \cdot 2005}{2}$. Since t is not an integer, there is no possible arrangement. For 2003 we can do this the following way.

1	2	$\cdots$	1002	1003	$\cdots$	2002	2003
1002	1003	$\cdots$	2003	1	$\cdots$	1000	1001
2003	2001	$\cdots$	1	2002	$\cdots$	4	2

In the first row we have all the numbers from 1 to 2003 in increasing order. In the second row we have the same but starting with 1002, and then again with 1 after 2003. In the third one we have first the odd numbers in decreasing order and then the even numbers in decreasing order. The column sum is always 3006. □

Solution 5.5 Let T be the number of pairs (E, p), where E is a student and p is a question that E got right.

- If all questions were easy, then each question p is in more than $\frac{n}{2}$ pairs. If all the students failed, then each student is in less than $\frac{k}{2}$ pairs. Hence, $k\frac{n}{2} < T < n\frac{k}{2}$, which is impossible.
- With an analogous argument we obtain that $k\lfloor\frac{n}{2}\rfloor \geq T \geq n\lceil\frac{k}{2}\rceil$, so we need that $T = \frac{nk}{2}$ and that both k and n be even. If in this case we number all students and questions, it is sufficient for each student to answer correctly all the questions with his same parity for the properties to hold. □

Solution 5.6 Suppose there are A_k classes with exactly k students. We count the number T of pairs (P, C) where C is a class of exactly k students and P is a pair of students in C. First, note that each pair P is being counted at most once (by the condition of the problem). Thus, $T \leq \binom{n}{2}$. However, each class C has exactly $\binom{k}{2}$ pairs. Thus,

$$\binom{k}{2}A_k \leq \binom{n}{2}.$$

If we solve for A_k and repeat this process for all k such that $2 \leq k \leq n$, we get $n - 1$ inequalities. By adding them up we obtain the following equation on the number M of classes

$$M = A_2 + \cdots + A_n \leq n(n-1)\left(\frac{1}{1 \cdot 2} + \frac{1}{2 \cdot 3} + \cdots + \frac{1}{(n-1) \cdot n}\right)$$

$$= n(n-1)\left(\left[1 - \frac{1}{2}\right] + \left[\frac{1}{2} - \frac{1}{3}\right] + \cdots + \left[\frac{1}{n-1} - \frac{1}{n}\right]\right)$$

$$= n(n-1)\left(1 - \frac{1}{n}\right)$$

$$= (n-1)^2. \qquad \square$$

Solution 5.7 Consider a group of $n - 1$ women and $m + 1$ men. We count how many sets of n persons there are according to how many men there are in each. If a set has at least k men, there are $\binom{m+1}{k}$ ways to choose the k men and $\binom{m+n-k}{n-k}$ ways to choose the rest. Also, each of these groups has at least one man. Thus, by inclusion-exclusion, we get

$$\sum_{k=1}^{n}(-1)^{k-1}\binom{m+1}{k}\binom{m+n-k}{n-k}$$

of these sets. Since there are $\binom{m+n}{n} = \binom{m+1}{0}\binom{m+n}{n-0}$ of them, we obtain

$$\binom{m+1}{0}\binom{m+n}{n-0} = \sum_{k=1}^{n}(-1)^{k-1}\binom{m+1}{k}\binom{m+n-k}{n-k}.$$

Moving everything to the left side of the equation we are done. □

Solution 5.8 Note that $\alpha\sigma\alpha^{-1}(\alpha(i)) = \alpha(\sigma(i))$. Thus $(\gamma_1, \gamma_2, \ldots, \gamma_k)$ is a cycle in the decomposition of σ if and only if $(\alpha(\gamma_1), \alpha(\gamma_2), \ldots, \alpha(\gamma_k))$ is a cycle in the decomposition of $\alpha\sigma\alpha^{-1}$. With this it is clear that σ and $\alpha\sigma\alpha^{-1}$ have the same cycle structure.

If σ and τ have the same cycle structure, let $(\gamma_1, \gamma_2, \ldots, \gamma_k)$ and $(\delta_1, \delta_2, \ldots, \delta_k)$ be two corresponding cycles in σ and τ, respectively. Define then $\alpha(\gamma_i) = \delta_i$. Since σ and τ have the same structure, α is defined on the entire set A and is a permutation. Also, $\alpha\sigma\alpha^{-1}(\delta_i) = \alpha\sigma\alpha^{-1}(\alpha(\gamma_i)) = \alpha\sigma(\gamma_i) = \alpha(\gamma_{i+1}) = \delta_{i+1} = \tau(\delta_i)$. Thus $\alpha\sigma\alpha^{-1} = \tau$, as we wanted. □

Solution 5.9 Suppose there are m associations, and let A be the one with the biggest number a of members. Denote by T the number of associations with at least one member of A. We know that each of the a members must be in another $k - 1$ associations. Note that any association different from A can have at most one member of A. Thus $T \geq (k - 1)a + 1$. Since $T \leq m$, this means that $a \leq \frac{m-1}{k-1}$. Hence, we have m associations, each with a number of members that goes from 1 to $\lfloor\frac{m-1}{k-1}\rfloor$. By the pigeonhole principle, at least k associations have the same number of members. □

Solution 5.10 Note that given a permutation f of $\{1, 2, \ldots, n\}$, $\sum_{i=1}^{n} T_f(i)$ is the number of numbers a such that $f^{12}(a) = a$. Thus $\sum_{f \in X} \sum_{i=1}^{n} T_f(i)$ is the number of pairs (f, a) such that f is a permutation of $\{1, 2, \ldots, n\}$ and a is a number such that $f^{12}(a) = a$.

We are going to fix a and count in how many pairs (f, a) it is. For this we have to count how many permutations there are such that $f^{12}(a) = a$. For this to happen a must be in a cycle of length 1, 2, 3, 4, 6 or 12 in the decomposition of f. Let us examine each case separately.

- If a is a fixed point, there are $(n-1)!$ ways to permute the rest.
- If a is in a cycle of length 2, there are $\binom{n-1}{1}$ ways to choose the other element of the cycle, 1! to form a cycle with it and a and $(n-2)!$ ways to permute the others, which gives $\binom{n-1}{1} \cdot 1! \cdot (n-2)! = (n-1)!$ permutations.
- If a is in a cycle of length 3, there are $\binom{n-1}{2} \cdot 2! \cdot (n-3)! = (n-1)!$ permutations.
- Analogously, if a is in a cycle of length 4, 6, or 12, then a is in $(n-1)!$ permutations.

Then a is in $6 \cdot (n-1)!$ pairs, so the total number of pairs is $6 \cdot n!$.

Another way to see that the number of permutations of $\{1, 2, \ldots, n\}$ (and use the different ways we have seen to represent a permutation) where a is in a cycle of length k is $(n-1)!$ is to order the other $n-1$ elements in a line. The first $k-1$ are going to be the ones in a's cycle with that order (taking a first). For the other $n-k$, compare the order you have with their original order; this should tell you how to permute them. For example, if $n = 10$ and $a = 7$, we give any permutation of the other 9 numbers, say $(1, 4, 5, 10, 9, 2, 3, 8, 6)$. The first 3 go in the cycle with 7 and give the cycle $(7, 1, 4, 5)$. The other 6 we compare with their original order, like in the next figure:

$$\begin{pmatrix} 2 & 3 & 6 & 8 & 9 & 10 \\ 10 & 9 & 2 & 3 & 8 & 6 \end{pmatrix}.$$

So the total permutation looks like

$$\begin{pmatrix} 1 & 2 & 3 & 4 & 5 & 6 & 7 & 8 & 9 & 10 \\ 4 & 10 & 9 & 5 & 7 & 2 & 1 & 3 & 8 & 6 \end{pmatrix},$$

with cycle structure $(1, 4, 5, 7)(2, 10, 6)(3, 9, 8)$. Since this function is clearly bijective, we are done. □

Solution 5.11 First Solution Consider the pairs (P, t) where P is a permutation of $\{1, 2, \ldots, n\}$ and t is a fixed point of P. Each element t of $\{1, 2, \ldots, n\}$ is in $(n-1)!$ pairs, since we only need to permute the other elements. If a permutation has k fixed points, it is in k pairs. Thus $n(n-1)! = \sum_{k=0}^{n} k P(k)$. □

Second Solution Let $P_n(k)$ be the number of permutations of $\{1, 2, \ldots, n\}$ with exactly k fixed points. If $k > n$ or $k < 0$, we put $P_n(k) = 0$. We solve the problem by induction on n. If $n = 1$, the assertion is true. Suppose that the assertion is true for n and we want to prove it for $n + 1$.

We show that

$$P_{n+1}(k) = P_n(k-1) + (n-k) P_n(k) + (k+1) P_n(k+1).$$

Let σ be a permutation of $\{1, 2, \ldots, n+1\}$ with exactly k fixed points. There are two cases, that $n + 1$ is a fixed point and that it is not. If it is a fixed point, there are $P_n(k-1)$ ways to permute the rest of the elements. If it is not a fixed point, we remove $n + 1$ and send $\sigma^{-1}(n + 1)$ to $\sigma(n + 1)$. Again we have two cases: that we get a new permutation of $\{1, 2, \ldots, n\}$ with k fixed points (there were no new fixed points added) or that there are $k + 1$ fixed points (if we are adding a fixed point). The first case is when $n + 1$ was in a cycle of length greater than or equal to 3 in the permutation and the second case is when it was in a cycle of length 2.

In the first case we are counting each permutation $n - k$ times, once for each element that is not fixed. This is because $n + 1$ could be placed in front of that element in the permutation to count it. In the second case we are counting each permutation $k + 1$ times, since $n + 1$ could be in a 2-cycle with any fixed point of the smaller permutation. This gives the equation above.

Thus, we have that

$$\sum_{k=0}^{n+1} k P_{n+1}(k) = \sum_{k=0}^{n+1} k \left[P_n(k - 1) + (n - k) P_n(k) + (k + 1) P_n(k + 1) \right]$$

$$= \sum_{k=0}^{n+1} k P_n(k - 1) + \sum_{k=0}^{n+1} k(n - k) P_n(k) + \sum_{k=0}^{n+1} k(k + 1) P_n(k + 1)$$

$$= \sum_{k=0}^{n+1} k P_n(k - 1) + \sum_{k=0}^{n+1} n \cdot k P_n(k) - \sum_{k=0}^{n+1} k^2 P_n(k)$$

$$+ \sum_{k=0}^{n+1} k(k + 1) P_n(k + 1)$$

$$= \sum_{k=1}^{n+1} k P_n(k - 1) + \sum_{k=0}^{n} n \cdot k P_n(k) - \sum_{k=0}^{n} k^2 P_n(k)$$

$$+ \sum_{k=0}^{n-1} k(k + 1) P_n(k + 1).$$

If we use the substitution $r = k - 1$ in the first sum, $r = k$ in the second and third and $r = k + 1$ in the fourth, we have

$$= \sum_{r=0}^{n} (r + 1) P_n(r) + n \sum_{r=0}^{n} r P_n(r) - \sum_{r=0}^{n} r^2 P_n(r) + \sum_{r=1}^{n} (r - 1) \cdot r P_n(r).$$

Note that by induction hypothesis the second sum is $n \cdot n!$. Also, in the fourth sum we can add the case $r = 0$ without changing the total. By expanding the expression above we obtain the following:

$$= \sum_{r=0}^{n} r P_n(r) + \sum_{r=0}^{n} P_n(r) + n \cdot n! - \sum_{r=0}^{n} r^2 P_n(r) + \sum_{r=0}^{n} r^2 P_n(r) - \sum_{r=0}^{n} r P_n(r)$$

$$= \sum_{r=0}^{n} P_n(r) + n \cdot n!.$$

But $\sum_{r=0}^{n} P_n(r) = n!$, since we are counting all the permutations, so we get $n! + n \cdot n! = (n+1)n! = (n+1)!$

This problem can also be solved by computing explicitly $P_n(k)$ using Example 1.4.6. However, the calculations that are needed are extremely long and tedious. $\square$

Solution 5.12 In each balanced sequence we can consider the zeros as delimiters. Thus, the ones are in $n+1$ possible places. In this way every sequence represents a list $(r_1, r_2, r_3, \ldots, r_{n+1})$, where r_i represents the number of ones in the i-th place. We have that $r_1 + r_2 + \cdots + r_{n+1} = n$. To every sequence we can assign the number between 1 and $n+1$ that is congruent to $r_1 + 2r_2 + 3r_3 + \cdots + (n+1)r_{n+1}$. That is, to the ones in the i-th position we assign the number i and then we compute the sum of all these numbers modulo $n+1$.

Note that each sequence $(r_1, r_2, r_3, \ldots, r_{n+1})$ with $r_1 + r_2 + \cdots + r_{n+1} = n$ has some j with $r_j > 0$. If we change a one in the j-th place to the i-th place we have changed the assigned number by $i - j$ modulo $n+1$. Since that i can be any number from 1 to $n+1$, the sequence has neighbors with all assigned numbers. Since there are in total $\binom{2n}{n}$ balanced sequences, there is a number that was assigned to at most $\frac{1}{n+1}\binom{2n}{n}$ sequences. If we consider S as the set of sequences that got this number, we are done. $\square$

Solution 5.13 We first show that every element of B is in exactly 2 of the A_i. For this, consider a set A_j. A_j has exactly one element in common with each of the other sets A_i. However, each of its $2n$ elements is in at least one other A_i. Since there are only $2n$ possible sets A_i with $i \neq j$, we cannot share one of its elements with more than one A_i. Thus, no element is in 3 or more of the A_i.

Now that we know this, suppose first that an assignment as the problem asks is possible. Let k denote the number of elements that have a 0 assigned and let T be the number of pairs (a, A_i) such that $a \in A_i$ and a has 0 assigned. For each A_i there are n pairs, so $T = n(2n+1)$. But each a is in 2 pairs, so $T = 2k$. Consequently, $2k = n(2n+1)$. Since $2n+1$ is odd, n must be even.

Now we show that for each even n we can make this assignment. For this, note that for each element of B there is a unique pair (A_i, A_j) of sets that have it as element. Therefore, it is sufficient to assign 0 or 1 to the pairs (i, j) with $1 \leq i < j \leq 2n+1$ so that each number is in exactly n pairs that were assigned 0. If this is possible, then to the number in $A_i \cap A_j$ we assign the same one as the number for the pair (i, j), and we are be done.

To do this assignment consider a regular polygon of $2n+1$ sides. We color its sides and diagonals in the following way. Consider a vertex a and split the other $2n$ vertices into n pairs that make parallel segments (those that have a in its orthogonal bisector). We start coloring the diagonals red and blue alternatingly so that the one closest to a is red. Note that if we do this for a given vertex, we are painting the same number of red segments as blue segments. The coloring is fixed if we rotate the polygon. Thus, for every two vertices v_1 and v_2, the number of red segments that use v_1 as endpoint and the number of red segments that use v_2 as endpoint are

the same. The same goes for blue segments. Since the total number of blue and red
segments is the same, at each vertex there must be the same number of red and blue
segments. We can assign to the pair (i, j) the number 0 if they define a red diagonal
and 1 if they define a blue one, which yields an assignment with the properties we
wanted.   □

Solution 5.14 Let T be the number of pairs (P, q), where q is a contestant and P
is a pair of judges that agree on q. The condition of the problem tells us that every
pair of judges P is in at most k of the pairs. Thus $\binom{n}{2}k \geq T$. But if we fix q, as n is
odd, q is in at least $\binom{\frac{n-1}{2}}{2} + \binom{\frac{n+1}{2}}{2}$ pairs. Thus $T \geq m\left[\binom{\frac{n-1}{2}}{2} + \binom{\frac{n+1}{2}}{2}\right]$. By combining
these two inequalities we obtain

$$\binom{n}{2}k \geq m\left[\binom{\frac{n-1}{2}}{2} + \binom{\frac{n+1}{2}}{2}\right] \Rightarrow$$

$$\frac{n(n-1)}{2}k \geq m\left[\frac{(\frac{n-1}{2})(\frac{n-3}{2})}{2} + \frac{(\frac{n+1}{2})(\frac{n-1}{2})}{2}\right] \Rightarrow$$

$$\frac{n(n-1)}{2}k \geq m\left[\frac{(\frac{n-1}{2})(\frac{2n-2}{2})}{2}\right] \Rightarrow$$

$$\frac{n}{2}k \geq m\left(\frac{\frac{1}{2}(n-1)}{2}\right) \Rightarrow$$

$$\frac{k}{m} \geq \frac{n-1}{2n}.$$ □

Solution 5.15 We say that $L = (L_1, L_2, \ldots, L_n)$ is a good list if for every $1 \leq i \leq n$,
L_i is a subset of i elements of $\{1, 2, \ldots, n\}$ and L_i is a subset of L_j for all $j \geq i$.
Note that these lists are like permutations[8] of $\{1, 2, \ldots, n\}$, where the permutation
tells us in which order we should add elements to the sets in the list to get the
next one. Let T denote the number of pairs (L, A_i) such that L is a good list and
A_i appears in L. By the condition of the problem, no two A_i's can appear in the
same list, so $T \leq n!$. Given an A_i with r elements, the lists it is in correspond
to permutations for which the first r elements are the ones in A_i. That is, we can
only permute the r elements of A_i at the beginning of the list and the other $n - r$
elements at the end of the list. If $f(A_i)$ is the number of lists A_i is in, then $f(A_i) =
r!(n - r)! \geq \lfloor\frac{n}{2}\rfloor!(n - \lfloor\frac{n}{2}\rfloor)!$. Thus

$$n! \geq T = \sum_{i=1}^{k} f(A_i) \Rightarrow$$

$$n! = \sum_{i=1}^{k} f(A_i) \geq \sum_{i=1}^{k} \left\lfloor\frac{n}{2}\right\rfloor!\left(n - \left\lfloor\frac{n}{2}\right\rfloor\right)! \Rightarrow$$

---

[8]In this context we are going to think of permutations as orderings of a list, and not as functions.

$$n! \geq k \cdot \left\lfloor \frac{n}{2} \right\rfloor! \left(n - \left\lfloor \frac{n}{2} \right\rfloor \right)! \quad \Rightarrow$$

$$\binom{n}{\lfloor \frac{n}{2} \rfloor} \geq k$$

as we wanted. Note that if the sets A_i are the all subsets with exactly $\lfloor \frac{n}{2} \rfloor$ elements we obtain a family of $\binom{n}{\lfloor \frac{n}{2} \rfloor}$ sets satisfying the condition of the problem, which shows that this number is the best possible.   □

Solution 5.16 Let p_0 be any person. Note that every time he plays, p_0 is put against 4 persons, and in total he must play against $n - 1$ persons. Thus $n - 1$ is divisible by 4. Let T be the number of pairs (P, Q) where P is a pair of persons and Q is a match in which the persons of P played in different teams. If a is the number of matches played, then, since in each match 16 pairs play each other, $T = 16a$. But by the condition of the problem $T = \binom{n}{2} = \frac{n(n-1)}{2}$. Thus $n(n-1)$ must be divisible by 32. Since we already know that n has to be odd, $n - 1$ must be divisible by 32. That is, $n = 32k + 1$ for some positive integer k.

We show that for every k we can make a tournament like this with $32k + 1$ persons. If $k = 1$, then we can name the persons $p_1, p_2, \ldots, p_{33}$ and order them in a cycle. Note that any two persons are at a distance of at most 16 in the cycle. Consider the teams $\{p_i, p_{i+1}, p_{i+2}, p_{i+3}\}$ and $\{p_{i+4}, p_{i+8}, p_{i+12}, p_{i+16}\}$ (the indices are mod 33). If we let these teams play against each other, then the 16 pairs that are in different teams have all the possible distances in the cycle (from 1 to 16). Thus if we move i we are not repeating pairs in different teams and every pair appears once.

Once we have the tournament for 33 persons, we make one for $32k + 1$ for every $k > 1$. For this take one person p_0 out and divide the rest into k groups of 32 persons each. If we include p_0 in any of these groups, we can make them play the tournament for 33 persons. Once we have done this for every group of 32, p_0 has played with each person once and each person has played with everyone else in his group once. Now divide each group of 32 into 8 teams of 4. To add the matches we were missing, it is sufficient for each of these new teams to play any other team that was not in their original group of 32 persons. Thus the answer is $n = 32k + 1$ for some positive integer k.   □

Solution 5.17   First Solution Let C_n^q be the number of prime necklaces with n pearls and q colors, and let L_n^q be the number of lists of pearls of q colors such that when we twine them to make a necklace they make a prime necklace (call them prime lists). We have that $nC_n^q = L_n^q$ since, given a prime necklace, when we "unroll" it starting from any pearl, we obtain a different prime list. We want to prove that $C_n^{q^n} = nC_{n^2}^q$, which is equivalent to $L_n^{q^n} = L_{n^2}^q$. Note that to each of the q^n colors we can assign a sequence of n pearls, each with one of q possible colors. Using this rule, we are assigning a list B of n^2 pearls of q different colors to each list A of $L_n^{q^n}$. First we show that B is a prime list. If this were not true, then there would be an integer $t > 1$ such that t divides n^2 and B is formed by taking t times a list of

$\frac{n^2}{t}$ pearls. Then, the same happens for any divisor of t, in particular $r = (t, n) > 1$. If we consider B as a list formed by repeating r times some list, the length of the repeating part is divisible by n, so the list A is also made of r identical parts. But then A would not be a prime list, a contradiction.

Now we have to verify that this way of assigning lists is bijective between these two sets. It is clear it is injective. To see that it must be surjective, we only have to note that each list with n^2 pearls of q colors can be divided into n lists of n pearls each, which gives us the inverse assignment. We only have to see that the "inverse" list A of a prime list B is also prime. If A were not prime, there would be an integer $r > 1$ such that A is formed by r times that list. In that case r satisfies the same for B, so B would not be prime. Thus the assignment must be surjective, which is the only thing that was left to prove. $\square$

Second Solution (with number theory) This solution requires more technical knowledge in number theory than we would like to ask for, but, since it gives a nice formula for the number of prime sequences, we present it here. For this solution, acquaintance with **Möbius inversion formula** is needed. Let μ be **Möbius function**, which is defined as

$$\mu(n) = \begin{cases} 1 & \text{if } n \text{ is } 1, \\ 0 & \text{if there is a prime } p \text{ such that } p^2 \text{ divides } n, \\ (-1)^r & \text{if } n \text{ is the product of } r \text{ primes.} \end{cases}$$

The Möbius inversion formula says that if f and g are functions such that for all n

$$f(n) = \sum_{d|n} g(d),$$

then for all n

$$g(n) = \sum_{d|n} \mu\left(\frac{n}{d}\right) g(d).$$

The proof is left to the interested reader. Going back to the problem, define the numbers L_m^k as before. Note that if a list is not prime, then it is a repetition of a prime list whose length is a divisor of n. Thus for al m, k,

$$k^m = \sum_{d|m} L_d^k,$$

which by the Möbius inversion yields a formula for the value of L_m^k. Namely,

$$L_m^k = \sum_{d|m} \mu\left(\frac{m}{d}\right) k^d.$$

If we substitute here $m = n^2$ and $k = q$ and then $m = n$ and $k = q^n$, we have to show that

$$\sum_{d|n^2} \mu\left(\frac{n^2}{d}\right) q^d = \sum_{d|n} \mu\left(\frac{n}{d}\right) q^{nd}.$$

Note that, if in the argument of μ there is a number divisible by the square of a prime, that term gives zero. The only terms that survive are those that give products of different primes (considering 1 as the product of 0 primes) in the argument of μ. If a product of different primes divides n^2, it also divides n. If d_1 and d_2 are divisors of n^2 and n, respectively, such that n^2/d_1 and n/d_2 are the same product of different primes, then $d_1 = nd_2$, which compensates the extra n on the exponent of q on the right-hand side of the equation. Thus, $L_{n^2}^q = L_n^{q^n}$, as we wanted.   □

Solution 5.18 Denote by A the set of sequences that are counted for N and by B the set of sequences that are counted for M. To each sequence in A we assign a sequence in B in the following way: whenever the state of the lamp $n + t$ was changed, we change the state of the lamp t instead. Given any sequence s_0 in B let us see how many sequences in A were assigned to s_0.

To obtain these sequences, we have to determine how many times the state of the lamp $n + t$ was changed for all $1 \le t \le n$ and the moments in which this happens. Let n_t be the number of times the state of the lamp t was changed in s_0. We have to choose a certain even number of steps of s_0 in which the state of the lamp t was changed and replace them by steps where the state of the lamp $n + t$ is changed. We know that given a set of n_t elements, the number of subsets with an even number of elements is 2^{n_t-1} (using Exercises 1.1.12 and 1.1.13). Thus there are 2^{n_t-1} possible choices for the lamp t. In total there must be $2^{n_1-1} \cdot 2^{n_2-1} \cdot \ldots \cdot 2^{n_n-1}$ sequences in A that go to s_0. However,

$$2^{n_1-1} \cdot 2^{n_2-1} \cdot \ldots \cdot 2^{n_n-1} = 2^{(n_1-1)+(n_1-1)+\cdots+(n_n-1)}$$
$$= 2^{(n_1+n_2+\cdots+n_n)-n} = 2^{k-n}.$$

Thus, $\frac{N}{M} = 2^{k-n}$.   □

Solution 5.19 Let T be the number of pairs (P, q) such that P is a pair of points of S and q is a point of S in the orthogonal bisector of P. Given a pair P, P is in at most two pairs of T, since no three points of S lie in a line. Given a point q, q is in at least $\binom{k}{2}$ pairs of T by the condition of the problem. Thus,

$$n\frac{k(k-1)}{2} \le T \le n(n-1) \quad \Rightarrow$$
$$k^2 - k \le 2n - 2 \quad \Rightarrow$$
$$(2k-1)^2 = 4k^2 - 4k + 1 \le 8n - 7 < 8n \quad \Rightarrow$$

$$2k - 1 < 2\sqrt{2n} \quad \Rightarrow$$
$$k < \sqrt{2n} + \frac{1}{2}.$$

$\qquad\qquad\qquad\qquad\qquad\qquad\qquad\qquad\qquad\qquad\qquad$ □

9.6   Solutions for Chap. 6

Solution 6.1

- Note that $\frac{n}{n!} = \frac{1}{(n-1)!}$. By the definition of the derivative, we are done.
- In $\exp(x)\exp(y)$ group together the terms of the type $x^r y^s$ with $r + s = n$. Their sum is precisely

$$\sum_{k=0}^{n} \frac{x^k}{k!} \frac{y^{n-k}}{(n-k)!} = \frac{1}{n!} \sum_{k=0}^{n} \binom{n}{k} x^k y^{n-k} = \frac{(x + y)^n}{n!},$$

as we wanted.
$\qquad\qquad\qquad\qquad\qquad\qquad\qquad\qquad\qquad\qquad\qquad$ □

Solution 6.2 Consider the generating function $f(x) = \frac{1}{1-x-x^2} = F_1 + F_2 x + F_3 x^2 + \cdots$. If we do the substitution $u = x + x^2$, as in Example 6.1.3, we have that

$$f(x) = 1 + (x + x^2) + (x + x^2)^2 + \cdots.$$

To find the coefficient of x^n, take a term of the form $(x + x^2)^k = x^k (1 + x)^k$ and find there the coefficient of x^n. By Newton's theorem, we have that $x^k(1 + x)^k = x^k \sum_{r=0}^{k} \binom{k}{r} x^r = \sum_{r=0}^{k} \binom{k}{r} x^{r+k}$. If $r + k = n$, the coefficient is $\binom{k}{n-k}$. Thus $F_{n+1} = \binom{n}{0} + \binom{n-1}{1} + \binom{n-2}{2} + \cdots$, as we wanted.
$\qquad\qquad\qquad\qquad\qquad\qquad\qquad\qquad\qquad\qquad\qquad$ □

Solution 6.3 Define $L_0 = -1$. With an argument analogous to the one used for the Fibonacci numbers, we obtain that if $L(x)$ is the generating function of the Lucas numbers, then

$$L(x) = \frac{1 - 3x}{x^2 + x - 1}$$
$$= \left(\frac{-1}{\phi^2}\right)\frac{1}{x - \frac{1}{\phi}} - (\phi^2)\frac{1}{x + \phi}$$
$$= \left(\frac{1}{\phi}\right)\frac{1}{1 - \phi x} - \phi \frac{1}{1 + \frac{x}{\phi}}.$$

Thus, $L_n = (\frac{1}{\phi})\phi^n - \phi(\frac{-1}{\phi})^n = \phi^{n-1} + (\frac{-1}{\phi})^{n-1}$.
$\qquad\qquad\qquad\qquad\qquad\qquad\qquad\qquad\qquad\qquad\qquad$ □

Solution 6.4 Let f be the generating function of the sequence. Then

$$\left(1 - 3x + 2x^2\right) f(x) = 1 + x^2 + 2x^2 + 4x^3 + \cdots$$
$$= 1 - x + x\left(1 + 2x + 4x^2 + 8x^3 + \cdots\right)$$
$$= 1 - x + x\left(\frac{1}{1 - 2x}\right).$$

Upon noting that $1 - 3x + 2x^2 = (1 - x)(1 - 2x)$, we obtain

$$f(x) = \frac{1}{1 - 2x} + \frac{x}{(1 - x)(1 - 2x)^2}.$$

We can decompose the second term into simple fractions as follows: $\frac{x}{(1-x)(1-2x)^2} = \frac{1}{(1-2x)^2} - \frac{2}{1-2x} + \frac{1}{1-x}$. Thus $f(x) = \frac{1}{(1-2x)^2} - \frac{1}{1-2x} + \frac{1}{1-x}$, so $a_n = (n + 1)2^n - 2^n + (-1)^n = n2^n + (-1)^n$, which is what we wanted to find. $\square$

Solution 6.5 Let $p = (1 + \frac{1}{1})(1 + \frac{1}{2})(1 + \frac{1}{3}) \cdots (1 + \frac{1}{n})$. To calculate this product we must choose in each term of the form $(1 + \frac{1}{k})$ either the 1 of the $\frac{1}{k}$ and multiply them. Then, consider the set of numbers k where we chose the part $\frac{1}{k}$. That is, we have to choose a subset of $\{1, 2, 3, \ldots, n\}$ and consider the inverse of the product of its elements. However, here we are also considering the empty product (when for each term we choose the 1). Thus, the number we want is $p - 1 = (\frac{2}{1})(\frac{3}{2})(\frac{4}{3}) \cdots (\frac{n+1}{n}) - 1 = (n + 1) - 1 = n$. $\square$

Solution 6.6 Number the points from 1 to $2n$. Let a_n be the number of ways to connect the points with $a_0 = 1$ without pairs of intersecting segments. Note that if we connect 1 with k, between them there must be an even number of points, so 1 must be connected to an even vertex. If it is connected with $2r$, then the $2r - 2$ points between 1 and $2r$ must be joined with points in that same section. The same holds for the other $2n - 2r$ points. Thus there are $a_{r-1}a_{n-r}$ ways to do this. We conclude that the sequence of a_n has the same recursion formula as c_n. This gives $a_n = c_n$ for every n. Note that, to match the points in an arbitrary way (regardless of the intersections), first we can choose n of them and then decide to which element of the other n we are pairing each of them. This can be done in $\binom{2n}{n}n!$ ways. However, we are counting each way to pair them 2^n times (in the set of the first n elements we chose we needed one element of each pair). Thus there is a total of $\frac{\binom{2n}{n}n!}{2^n}$ ways to do this. It follows that the probability we were searching for is

$$\frac{c_n}{\frac{\binom{2n}{n}n!}{2^n}} = \frac{2^n c_n}{\binom{2n}{n}n!} = \frac{2^n \binom{2n}{n}}{\binom{2n}{n}n!(n + 1)} = \frac{2^n}{(n + 1)!}.$$

$\square$

Solution 6.7 Consider an $n \times n$ board and call a path "good" if it starts on the lower left corner, always goes up or to the right, never goes above the main diagonal and has length n. That is, if we assign coordinates to the vertices of the board, the path starts at the point $(0,0)$, goes to a point of the form $(i, n-i)$ for some i and remains in the region $H = \{(x,y) \mid x \geq y\}$. Clearly every sequence we are trying to count corresponds to one of these paths. Let T denote the number of paths of length n that end in H but do go above the main diagonal at some point and have their endpoint in H. For each of these paths, consider the first moment it is above the main diagonal. Then, using the same reflection as in the proof of the closed-form formula for the Catalan numbers we obtain a new path assigned to it. Notice that the new path has to end in a point of the form $(r, n-r)$ with $0 \leq r \leq \lfloor \frac{n}{2} \rfloor - 1$ (to see this clearly, it is convenient to treat the cases where n is even and n is odd separately). Also, for every path that ends in one of these points we can consider the first moment it is above the main diagonal and do this reflection to obtain a path counted in T. Thus

$$T = \binom{n}{0} + \binom{n}{1} + \cdots + \binom{n}{\lfloor \frac{n}{2} \rfloor - 1}.$$

Also, $A + T$ is the number of paths that end up in a point of the form $(r, n-r)$ with $r \geq \lceil \frac{n}{2} \rceil$. Thus

$$A + T = \binom{n}{n} + \binom{n}{n-1} + \cdots + \binom{n}{\lceil \frac{n}{2} \rceil} = \binom{n}{0} + \binom{n}{1} + \cdots + \binom{n}{\lfloor \frac{n}{2} \rfloor}.$$

By subtracting one equation from the other we obtain $A = \binom{n}{\lfloor \frac{n}{2} \rfloor}$, as we wanted. □

Solution 6.8 **First Solution (with generating functions)** Let $a_t = \binom{2t}{t}$ and let $F(x)$ be the generating function of the sequence $(a_0, a_1, a_2, \ldots)$. What we want to prove is that $F^2(x) = \frac{1}{1-4x}$, i.e., that $F(x) = (1-4x)^{-\frac{1}{2}}$.

We can note that $(1-4x)^{-\frac{1}{2}}$ is the generating function of the sequence $(b_0, b_1, b_2, \ldots)$ where $b_k = (-4)^k \binom{-\frac{1}{2}}{k}$,

$$(-4)^k \binom{-\frac{1}{2}}{k} = (-4)^k \frac{(-\frac{1}{2})(-\frac{1}{2}-1)(-\frac{1}{2}-2)\cdots(-\frac{1}{2}-k+1)}{k!}$$

$$= 4^k \frac{(\frac{1}{2})(\frac{1}{2}+1)(\frac{1}{2}+2)\cdots(\frac{1}{2}+k-1)}{k!}$$

$$= 2^k \frac{1 \cdot 3 \cdot 5 \cdot \ldots \cdot (2k-1)}{k!}$$

$$= \frac{(2^k k!)[1 \cdot 3 \cdot 5 \cdot \ldots \cdot (2k-1)]}{(k!)^2} = \frac{(2k)!}{(k!)^2} = \binom{2k}{k}.$$

□

Second Solution (without generating functions) Consider the sequences of $2k + 1$ letters a or b in which there are more letters a than b. Clearly half of the 2^{2k+1} possible sequences work, so there are 4^k sequences we are interested in. Let us start writing the elements of the sequence one by one, and denote by $2r$ the last place where there is the same number of letters a and letters b written up to that point ($r = 0$ is allowed). Up to that moment there are $\binom{2r}{r}$ sequences possible. The next letter must be an a, since otherwise we can find a point later where there is the same number of a's and b's written up to that point. Consider the sequence of $2r - 2t$ letters after that a. If at any point there are more letters b than a in that sequence, one contradicts the maximality of $2r$ in the original sequence. By Problem 6.7, there are $\binom{2r-2t}{r-t}$ possibilities for this part of the sequence. Thus

$$\binom{2k}{k}\binom{0}{0} + \binom{2k-2}{k-1}\binom{2}{1} + \binom{2k-4}{k-2}\binom{4}{2} + \cdots + \binom{0}{0}\binom{2k}{k} = 4^k. \qquad \square$$

Solution 6.9 Let $b(n)$ be the number of sequences of size n with an even number of ones, considering $b_0 = 1$. Let $A(x)$ and $B(x)$ be the generating functions of the sequences a_n and b_n, respectively. Note that in total there are 4^n sequences of size n, so $a_n + b_n = 4^n$. This formula works even if $n = 0$, so $A(x) + B(x) = \frac{1}{1-4x}$. Also, $a_{n+1} = 3a_n + b_n$ since every sequence of $n + 1$ digits with an odd number of 1 can have a 1 as the last digit or something else. If it is a 1, there are b_n possibilities for the first n digits. If it is not a 1, there are a_n possibilities for the first n digits. Adding up we get the previous equality. Thus $A(x) = x[3A(x) + B(x)]$. Using these two equations we can solve for $A(x)$ and obtain

$$A(x) = \frac{x}{(1-2x)(1-4x)} = \frac{1}{2}\left(\frac{1}{1-4x}\right) - \frac{1}{2}\left(\frac{1}{1-2x}\right).$$

Thus $a_n = \frac{4^n - 2^n}{2}$. $\qquad \square$

Solution 6.10 Using what was done in Example 6.4.2 we have that the generating function of the sequence $(0^2, 1^2, 2^2, \ldots)$ is $f(x) = \frac{x(1+x)}{(1-x)^3}$. The generating function associated with the cubes is then

$$h(x) = xf'(x) = \frac{x^3 + 4x^2 + x}{(1-x)^4}.$$

It follows that the generating function associated with the sum of the first n cubes is $\frac{x^3+4x^2+x}{(1-x)^5} = \frac{x^3}{(1-x)^5} + \frac{4x^2}{(1-x)^5} + \frac{x}{(1-x)^5}$. The sum of the first n cubes is

$$\binom{n+1}{4} + 4\binom{n+2}{4} + \binom{n+3}{4}$$

$$= \frac{(n+1)n(n-1)(n-2) + 4(n+2)(n+1)n(n-1) + (n+3)(n+2)(n+1)n}{4!}$$

$$= \frac{n(n+1)[(n-1)(n-2)+4(n+2)(n-1)+(n+3)(n+2)]}{4!}$$

$$= \frac{n(n+1)[6n^2+6n]}{4!} = \left[\frac{n(n+1)}{2}\right]^2.$$ □

Solution 6.11 Since three pins (P) or two brackets (B) cannot lie in a row, they cannot do so on an individual brick. This means that there are only three different kind of bricks. These are $A = (PBPP)$, $B = (PPBP)$, $C = (BPB)$. Let a_n, b_n and c_n denote the number of sequences of n bricks that end with a brick of type A, B, and C, respectively. Note that, by the condition of the problem, $a_{n+1} = b_n + c_n$, $b_{n+1} = c_n$, $c_{n+1} = a_n + b_n$ and $a_1 = b_1 = c_1 = 1$. We are interested in $d_n = a_n + b_n + c_n$. This gives

$$d_{n+1} = d_n + b_n + c_n = d_n + c_{n-1} + a_{n-1} + b_{n-1} = d_n + d_{n-1},$$

that is, the same recurrence relation as for the Fibonacci numbers. Note that $d_1 = 3$ and $d_2 = 5$. Thus $d_n = F_{n+3}$.

With the recursion relations we can also obtain formulas for $A(x)$, $B(x)$, $C(x)$, the generating functions of the sequences (a_n), (b_n), (c_n), and extract closed formulas for these sequences.   □

Solution 6.12 Let us try to find the coefficient of x^k in that product. For this, we have to choose in some of the terms the monomial x^{2^i}. Let $i_1, i_2, \ldots, i_n$ be the terms where monomials $x^{2^{i_r}}$ were chosen. Thus we have that $k = 2^{i_1} + 2^{i_2} + \cdots + 2^{i_n}$. What we are really doing is writing k in base 2, which can be done in a unique way. This means that the coefficient of x^k is 1, as we wanted.   □

Solution 6.13 First Solution Let $f(x)$ be the generating function that has coefficients 1 in the elements of F. Notice that the coefficient of x^k in $f(x)f(x^2)$ is the number of ways to write k as $a + 2b$ with a and b in F. Thus what we want is $f(x)f(x^2) = \frac{1}{1-x}$. By Problem 6.12, we have that $\frac{1}{1-x} = (1+x)(1+x^2)(1+x^4)(1+x^8)\cdots$, so $f(x) = (1+x)(1+x^4)(1+x^{16})\cdots$ works. Note that in this product there is no term with coefficient bigger than 1, so it does define a set.   □

Second Solution Recall that every number can be written in a unique way as a sum of different powers of 2. Let F be the set of positive integers that can be written as a sum of different even powers of 2 (be careful, by even we mean even in the exponent). Since every positive integer can be written in a unique way as sum of powers of 2, we can split the even powers from the odd powers. If the sum of the even powers is a and the sum of the odd power is $2b$, then a and b are in F and $a + 2b$ is the only way to write the number as the problem asks.   □

Solution 6.14 Using the recursive formula we obtain that the generating function for the sequence a_n is

$$A(x) = \frac{1-x}{1-3x+4x^2}$$

$$= \frac{1-x}{(1+x)(1-4x)}$$

$$= \frac{3}{5}\left(\frac{1}{1-4x}\right) + \frac{2}{5}\left(\frac{1}{1+x}\right).$$

It follows that

$$a_n = \frac{3 \cdot 4^n + 2 \cdot (-1)^n}{5}. \qquad \square$$

Solution 6.15 Consider $P(x) = x + 2x^2 + 3x^3 + \cdots$. To obtain a term with coefficient 31 in $P(x)^4$ we have to choose 4 terms ix^i, jx^j, kx^k, lx^l of $P(x)$ such that $i + j + k + l = 31$ and multiply them. This means that they add exactly $ijkl$ to the coefficient x^{31}. However, by what was done in Example 6.4.2 we have that $P(x) = \frac{x}{(1-x)^2}$. Thus $P(x)^4 = \frac{x^4}{(1-x)^8}$. Using what was done in Exercise 6.1.4 we have that this is the generating function of $\binom{n+3}{7}$. Thus the sum we want is $\binom{34}{7} = \frac{34 \cdot 33 \cdots 28}{7!}$. Since in the numerator there is a 31 and there is none in the denominator, the number is a multiple of 31. $\qquad \square$

Solution 6.16 We say a coloring of S is good if it satisfies the conditions of the problem. Let c_n be the number of good colorings where n is not painted, a_n the number of good coloring where n is white, and b_n the number of good colorings where n is black. Note that if n is not colored, the other $n - 1$ numbers need to obey no extra conditions, so $c_n = a_{n-1} + b_{n-1} + c_{n-1}$. If n is white, then $n - 1$ cannot be white, so $a_n = c_{n-1} + b_{n-1}$. We could say something similar of b_n, but note that any coloring of a_n after swapping the colors is a coloring of b_n and vice versa, so $a_n = b_n$. Thus $c_n = 2a_{n-1} + c_{n-1}$ and $a_n = a_{n-1} + c_{n-1}$. We can consider $a_0 = 0$ and $c_0 = 1$ so that their recursion formula is true even for $n = 1$.

Thus if $C(x)$ and $A(x)$ are the generating functions of these sequences, we have that $C(x) = x[2A(x) + C(x)] + 1$ and $A(x) = x[A(x) + C(x)]$. Solving for $C(x)$, we obtain

$$C(x) = \frac{x-1}{x^2+2x-1}, \qquad A(x) = \frac{-x}{x^2+2x-1}.$$

The roots of the polynomial $x^2 + 2x - 1$ are $\alpha = -1 + \sqrt{2}$ and $\beta = -1 - \sqrt{2}$. Thus $C(x) = \frac{1-\sqrt{2}}{2}\left(\frac{1}{x-\alpha}\right) + \frac{1+\sqrt{2}}{2}\left(\frac{1}{x-\beta}\right)$, which is equal to $\frac{1}{2}\left(\frac{1}{1+\beta x} + \frac{1}{1+\alpha x}\right)$. This gives the following

$$c_n = \frac{(\sqrt{2}+1)^n + (1-\sqrt{2})^n}{2}.$$

But note that the number of ways to color the first $n + 1$ numbers without painting the last one is the same as the number of ways to color the first n numbers without any extra conditions. Thus the number we want is

$$\frac{(\sqrt{2}+1)^{n+1} + (1 - \sqrt{2})^{n+1}}{2}.$$

Note: The problem can be finished without this last observation by solving also for $A(x)$ and obtaining $a_n = \frac{\sqrt{2}(1+\sqrt{2})^n - \sqrt{2}(1-\sqrt{2})^n}{4}$. After this the number we want is clearly $c_n + 2a_n$ and adding up we get the same result.   □

Solution 6.17 Note that the problem is the same as if it was in a $2 \times (n - 1)$ board with the path moving along its edges. Then the path encloses a section to its left, as in the following picture:

Note that every region as the one in the figure is associated with a unique path, so we can count the regions instead of the paths. The important properties of these regions are that we can get from one square to any other by moving by squares that share a side and that the whole region is attached to the left side. Let d_n be the number of these regions in a $2 \times n$ board where the last column is not painted. Let a_n be the number of these regions where in the last column only the top square is painted, b_n is the same with only the lower square painted and c_n with both squares painted.

Note that if we color in a $2 \times (n + 1)$ board so that in the last column only the top square is painted, then it also must be painted in the previous column. This translates to $a_{n+1} = a_n + c_n$. The same way we obtain that $c_{n+1} = a_n + b_n + c_n$ and that $d_{n+1} = d_n + c_n + a_n + b_n$. Since every coloring can be inverted (with respect to the horizontal direction), we have $a_n = b_n$. If we also define $a_0 = d_0 = 0$ and $c_0 = 1$[9] we ensure that the recursion formulas also work for $n = 0$.

Consider the generating functions $A(x)$, $C(x)$, $D(x)$ of the sequences a_n, c_n, d_n, respectively. What the recursion formulas say is that

$$A(x) = x\big(A(x) + C(x)\big),$$
$$C(x) = x\big(2A(x) + C(x)\big) + 1,$$
$$D(x) = x\big(D(x) + 2A(x) + C(x)\big).$$

---

[9]The reason why this makes sense is that we can attach a dummy column to the left. The condition of the region being attached to the left can be thought of as having the two squares of the dummy column to be painted (this is due to the fact that the region must be connected, which gives a meaning to the terms with index 0).

Solving for $D(x)$ we obtain $D(x) = \frac{x^2+x}{(1-2x-x^2)(1-x)}$. The roots of $1 - 2x - x^2$ are $\alpha = -1 - \sqrt{2}$ and $\beta = -1 + \sqrt{2}$. We have that

$$D(x) = \frac{x^2+x}{(1-2x-x^2)(1-x)} = \left(\frac{-1}{2\sqrt{2}}\right)\frac{1}{x-\alpha} + \left(\frac{1}{2\sqrt{2}}\right)\frac{1}{x-\beta} + \frac{1}{x-1}$$

$$= \left(\frac{1+\sqrt{2}}{2\sqrt{2}}\right)\frac{1}{1-(1+\sqrt{2})x} - \left(\frac{1-\sqrt{2}}{2\sqrt{2}}\right)\frac{1}{1-(1-\sqrt{2})x} - \frac{1}{1-x}.$$

It follows that

$$d_n = \frac{(1+\sqrt{2})(1+\sqrt{2})^n - (1-\sqrt{2})(1-\sqrt{2})^n}{2\sqrt{2}} - 1$$

$$= \frac{(1+\sqrt{2})^{n+1} - (1-\sqrt{2})^{n+1}}{2\sqrt{2}} - 1.$$

However, the number of regions of a $2 \times n$ board that do not use the last column is equal to the number of regions in a $2 \times (n-1)$ board, which we had seen is equal to $R(3, n)$. Thus

$$R(3, n) = \frac{(1+\sqrt{2})^{n+1} - (1-\sqrt{2})^{n+1}}{2\sqrt{2}} - 1. \qquad \square$$

Solution 6.18 First color the vertices of the octagon alternatingly black and white. In each jump the frog gets to a vertex of different color, so it needs to get to the opposite vertex in an even number of jumps.

After this, label the vertices from 0 to 7 (with the frog starting at 0) and let x_n be the number of ways to be at vertex 0 after $2n$ jumps without having passed by vertex 4, y_n the number of ways to be at vertex 2 after $2n$ jumps without having passed by vertex 4 and z_n the same but with vertex 6. Let $X(x)$, $Y(x)$ and $Z(x)$ be the generating functions of the sequences x_n, y_n, and z_n, respectively.

We have that $a_{2n} = y_{n-1} + z_{n-1}$ since two jumps before getting to vertex 4 for the first time the frog has to be at vertex 2 or 6. Note that, in order to get to vertex 0, two jumps before the frog could be at vertex 0, 2 or 6. If the frog was at vertex 0 it could go up and down or down and up, and so $x_{n+1} = y_n + 2x_n + z_n$. With an analogous reasoning we obtain $y_{n+1} = x_n + 2y_n$ and $z_{n+1} = x_n + 2z_n$. Using this and the fact that $x_0 = 1$ and $y_0 = z_0 = 0$ we have that, in terms of the generating functions, the following holds:

$$X(x) = x[Y(x) + 2X(x)] + 1,$$
$$Y(x) = x[X(x) + 2Y(x)],$$
$$Z(x) = x[X(x) + 2Z(x)].$$

By simple algebraic manipulations we obtain that $Y(x) = Z(x) = \frac{x}{2x^2-4x+1}$, so if $A(x)$ is the generating function of the sequence a_{2n} we have that

$$A(x) = \frac{2x^2}{2x^2 - 4x + 1} = x\left(\frac{2x}{2x^2 - 4x + 1}\right)$$
$$= x\left(\frac{x}{(x - \alpha)(x - \beta)}\right),$$

with $\alpha = \frac{2+\sqrt{2}}{2}$ and $\beta = \frac{2-\sqrt{2}}{2}$. Therefore,

$$A(x) = \frac{x}{\sqrt{2}}\left(\frac{-1}{1 - \frac{x}{\alpha}} + \frac{1}{1 - \frac{x}{\beta}}\right).$$

But since $\alpha\beta = \frac{1}{2}$ we have

$$A(x) = \frac{x}{\sqrt{2}}\left(\frac{-1}{1 - 2\beta x} + \frac{1}{1 - 2\alpha x}\right).$$

We know that $\frac{1}{1-2\alpha x}$ is the generating function of $(2 + \sqrt{2})^n$ and $\frac{-1}{1-2\beta x}$ is the one of $-(2 - \sqrt{2})^n$. Taking into consideration the factor $\frac{x}{\sqrt{2}}$, we obtain

$$a_{2n} = \frac{(2 + \sqrt{2})^{n-1} - (2 - \sqrt{2})^{n-1}}{\sqrt{2}}.$$ □

9.7   Solutions for Chap. 7

Solution 7.1 Consider a partition of the primes that divide l. Clearly we can assign a factorization of l by taking the factors to be the product of the numbers in each set of the partition. Since this assignment is bijective, we are done. □

Solution 7.2 By Example 1.1.5, $(a+b)_n$ is the number of ordered lists of n different numbers from a set of $a + b$ possible choices. Say a elements are red and b are blue. Another way to count such lists is by how many elements are going to be red. If we want k red elements, then we first choose their place in the list, which can be done in $\binom{n}{k}$ ways. After that, we only have to choose the red list (there are $(a)_k$ ways) and the blue list (there are $(b)_{n-k}$ ways). Thus $(a + b)_n = \sum_{k=0}^{n} \binom{n}{k}(a)_k(b)_{n-k}$.[10] □

Solution 7.3 For each self-conjugate partition we can consider its Ferrer diagram. To each square in the main diagonal assign the squares on its right and above it. Consider the sets formed by an element of the main diagonal and the squares assigned to it. Since the numbers of such squares (those above and those to the right)

---

[10]It is possible to prove this fact using only Newton's Theorem and the formulas relating the Stirling numbers of the first and second kind and these polynomials. However, this is rather onerous and should not be tried by the weak of heart.

are the same, each set has an odd number of elements. Also, no two of them can have the same number of squares since the diagram goes in a decreasing way. With this we have that the assignment is clearly bijective and goes between the self-conjugate partitions of n and the partitions of n where the parts are odd and pairwise distinct.

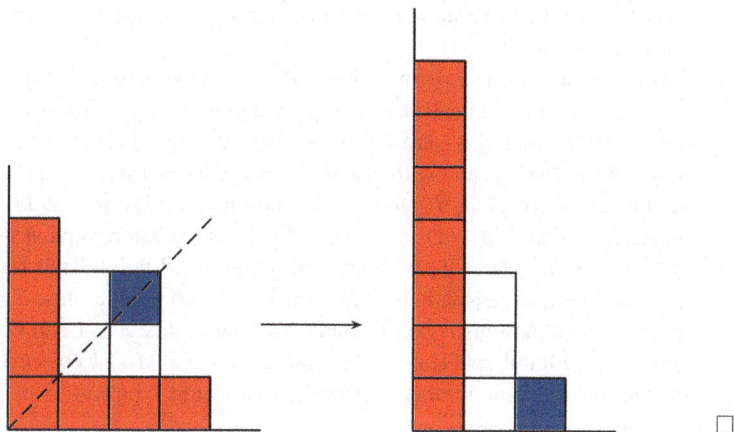

Solution 7.4 First Solution We will construct the sets of m elements one by one. For the first m there are $\binom{mn}{m}$ possibilities, then there are $\binom{(n-1)m}{m}$ ways of choosing the next m, then there are $\binom{(n-2)m}{m}$ and so on until get $\binom{m}{m}$ ways of choosing the last m. Note that each partition we wanted is being counted $n!$ times, since it depends on the order in which we choose the sets. Then the total number of ways is

$$\frac{1}{n!}\binom{mn}{m}\binom{(n-1)m}{m}\binom{(n-2)m}{m}\cdots\binom{m}{m}$$
$$= \frac{1}{n!}\frac{(mn)!}{m![(n-1)m]!}\frac{[(n-1)m]!}{m![(n-2)m]!}\frac{[(n-2)m]!}{m![(n-3)m]!}\cdots\frac{m!}{m!0!}.$$

Note that the numbers of the type $[(n-k)m]!$ cancel out, so we are left with

$$\frac{(mn)!}{(m!)^n n!}.$$ □

Second Solution Write the mn elements in a list. There are $(mn)!$ ways to do this. After that we are going to make the first set with the first m elements in the list, the second set with the next m elements, and so on. Note that each partition is being counted $m!$ times for each ordering the m sets of each of its sets and $n!$ times for each way to order the sets in the list. Thus the total number of partitions is

$$\frac{(mn)!}{(m!)^n n!}.$$ □

Solution 7.5

- First count the number of partitions of $[n]$ with exactly k cycles that have $n - t$ fixed points. For this we choose the t points that are not going to be fixed, and we want to permute them in such a way that they have exactly $k - (n - t)$ cycles and no fixed points. There are $d(t, t + k - n)$ ways to do this, so there are $\binom{n}{t} d(t, t + k - n)$ permutations with k cycles and $n - t$ fixed points. Since $(-1)^{n+k} s(n, k)$ is the number of permutations with exactly k cycles, by adding up we are done.

- Let σ be one of these permutations. There are two cases, that $\sigma(\sigma(n + 1))$ is $n + 1$ and that it is not. If it is $n + 1$, we can remove $n + 1$ and $\sigma(n + 1)$ from the permutation and we are left with a permutation of a list of $n - 1$ elements without fixed points and with exactly $k - 1$ cycles. Since $\sigma(n+1)$ has n options, in this case there are $n \cdot d(n - 1, k - 1)$ permutations. If $\sigma(\sigma(n+1))$ is not $n+1$, that means that $\sigma(n + 1) \neq \sigma^{-1}(n + 1)$. Thus we can remove $n + 1$ and send $\sigma^{-1}(n + 1)$ to $\sigma(n + 1)$ without adding any fixed point. That leaves us with a permutation of $[n]$ with exactly k cycles and no fixed points. Note that each of these permutations is being counted n times, since $n + 1$ can be added after any of the n numbers. Thus in this case there are $n \cdot d(n, k)$ permutations. This means that the total number of permutations of this type is $n \cdot d(n - 1, k - 1) + n \cdot d(n, k)$. □

Solution 7.6 We are going to count how many partitions there are such that the set in which n lies is missing exactly k elements of $[n]$. For this, choose the k elements that n is not going to be with and a partition of them. There are $\binom{n-1}{k} T(k)$ ways to do this, so we obtain the desired result. □

Solution 7.7 Let $(n_1, n_2, \ldots, n_k)$ some partition of m. Consider the term $\frac{1}{1-x^n} = 1 + x^n + x^{2n} + x^{3n} + \cdots$ in the product. In that term we are going to choose the monomial $x^{r_n \cdot n}$, where r_n is the number of times n appears in the partition. It is clear that when we do this with each term, we obtain a 1 to add to the coefficient of the monomial x^m in the product. Since every product of terms of this type that adds a 1 to the coefficient of x^m induces in the same way a partition of m, the final coefficient must be $p(m)$, as we wanted to prove. □

Solution 7.8 First Solution (with generating functions) Let $q(m)$ be the number of partitions of m whose parts are pairwise different and $q^*(m)$ the number of partitions of m whose parts are all odd. With an argument similar to the one used in Problem 7.7, with $q(m)$ we are only interested in the case when $r_n = 1$ or $r_n = 0$ for each n. Thus its generating function is $\prod_{n \geq 1}(1 + x^n)$. With $q^*(m)$ we are interested in doing this reasoning only when n is odd, so we get the generating function $\prod_{k \geq 1} \frac{1}{1-x^{2k-1}}$. Note that

$$\prod_{n \geq 1}\left(1 + x^n\right) = \prod_{n \geq 1} \frac{1 - x^{2n}}{1 - x^n}.$$

Thus in the numerator we have all the terms of the form $1 - x^t$ with t even and in the denominator we have all the term of the type $1 - x^t$. When they cancel out, only the odd terms remain in the denominator, as we wanted.   □

Second Solution (with bijective functions) Let $(n_1, n_2, \ldots, n_k)$ be a partition of n such that all its parts are different. Then each n_i can be written as $2^{a_i} b_i$, where b_i is odd. We remove each of the n_i and replace them by their b_i but 2^{a_i} times. In this way we obtain a partition of n with only odd parts. We will see that the function defined in this manner is bijective. For this it suffices to show that there is an inverse function. That is, we are given a partition $(n_1, n_2, \ldots, n_k)$ of n with only odd parts and we want to construct a partition with only different parts. For this suppose that b_i appears C_i times. We know that C_i can be written in a unique way as a sum of different powers of 2, say as $2^{a_{i,1}} + 2^{a_{i,2}} + \cdots + 2^{a_{i,s}}$. We remove all the b_i and write the numbers $2^{a_{i,1}} b_i, 2^{a_{i,2}} b_i, \ldots, 2^{a_{i,s}} b_i$. This yields a partition of n in which all parts are different. It is clear that the two functions constructed are inverse of each other, so they are both bijective, and we are done.   □

Solution 7.9 Consider that the first set is the one with the element 1. Then 2 must be in some other set, call it the second set. 3 cannot be in the second set, so it has to be in the first or third set (it has 2 options). In the same way, once 3 has been placed, 4 has two options, and so on. Note that when we do this counting we are considering the case where a set has all the odd elements, another set has all the even ones, and the last one is empty. Since this is the only case we have to eliminate, there are $2^{n-2} - 1$ partitions of the type we want.   □

Solution 7.10 We count the partitions of $[n + 1]$ by how many elements of $[n + 1]$ are not in the same set of the partition as $n + 1$. Since there must be another m sets, we need a least m elements not to be with $n + 1$. Once we have chosen the number t of elements that are not going to $n + 1$, we have to choose these elements and partition them into t non-empty subsets. This can be done in $\binom{n}{t} S(t, m)$ ways. Thus $\binom{n}{m} S(m, m) + \binom{n}{m+1} S(m + 1, m) + \cdots + \binom{n}{n} S(n, m) = S(n + 1, m + 1)$.   □

Solution 7.11 First note that the number of surjective functions from $[n]$ to $[m]$ is $m! S(n, m)$. This is because if we consider a partition of $[n]$ into m parts and then assign to each part an element of $[m]$ we are really constructing a surjective function from $[n]$ to $[m]$. Also, the number of functions from $[n]$ to $[n]$ is n^n.

   We count the functions from $[n]$ to $[n]$ by the size of their image. If their image has m elements, there are $\binom{n}{m}$ ways to choose the elements in the image. Thus there are $\binom{n}{m} m! S(n, m)$ functions from $[n]$ to $[n]$ such that their image has exactly m elements. This means that

$$n^n = \sum_{m=0}^{n} \binom{n}{m} m! S(n, m).$$

Simple algebraic manipulations of this equation give the desired result.   □

Notation

Throughout the book the following notation is used:

$a \in A$	The element a belongs to the set A
$A \subset B$	A is a subset of B
$\{a \in A \mid \psi(a)\}$	The elements of A that satisfy property ψ
$\{a_1, a_2, \ldots, a_r\}$	The set consisting of the elements $a_1, a_2, \ldots, a_r$
$A \cap B$	The intersection of the sets A and B
$A \cup B$	The union of the sets A and B
$\mathcal{P}(A)$	The power set of A
$\bigcap C$	The intersection of all sets of C
$\bigcup C$	The union of all sets of C
$\emptyset$	The empty set
$\|A\|$	The number of elements of set A
$\|y\|$	The absolute value of the real number y
$x < y$	x is smaller than y
$x \leq y$	x is smaller than or equal to y
$x \neq y$	x is not equal to y
$\lceil x \rceil$	The smallest integer greater than or equal to x
$\lfloor x \rfloor$	The greatest integer smaller than or equal to x
$a \equiv b \pmod{c}$	a is congruent to b modulo c, $a - b$ is divisible by c
$\sum_{k=i}^{j} \psi(k)$	$\psi(i) + \psi(i+1) + \psi(i+2) + \cdots + \psi(j)$
$\prod_{k=i}^{j} \psi(k)$	$\psi(i) \cdot \psi(i+1) \cdot \psi(i+2) \cdot \ldots \cdot \psi(j)$
$n!$	$1 \cdot 2 \cdot 3 \cdot \ldots \cdot n$
$\binom{n}{k}$	$\frac{n!}{k!(n-k)!}$
F_n	The Fibonacci numbers
L_n	The Lucas numbers
c_n	The Catalan numbers
$s(n, k)$	The Stirling numbers of the first kind
$c(n, k)$	The unsigned Stirling numbers of the first kind
$S(n, k)$	The Stirling numbers of the second kind
$r(l, s)$	The Ramsey numbers
$m \times n$	The dimensions of a board of m rows and n columns
$v(G)$	The set of vertices of a graph G
$e(G)$	The set of edges of a graph G

P. Soberón, *Problem-Solving Methods in Combinatorics*,
DOI 10.1007/978-3-0348-0597-1, © Springer Basel 2013

$d(v)$	The degree of vertex v
N_v	The connected component of v
$\Gamma(S)$	The vertices adjacent to some vertex of S
$f : A \longrightarrow B$	f is a function from A to B
$f[A]$	The image of the set A under f
$f(a)$	The element assigned to a by f
$\sigma \circ \tau$	The composition of σ with τ
σ^{-1}	The inverse permutation of σ
σ_A	σ restricted to A (defined only if $\sigma[A] = A$)
$(\gamma_1, \gamma_2, \ldots, \gamma_k)$	The permutation that sends each γ_i to γ_{i+1} and γ_k to γ_1
$(a_0, a_1, a_2, \ldots)$	The sequence $a_0, a_1, \ldots$
$f'(x)$	The derivative of the generating function f
$(t)_n$	$t(t-1)\ldots(t-n+1)$, if $n \geq 1$, $(t)_0 = 1$

When making references to problems the following abbreviations were used

IMO	International Mathematical Olympiad
OIM	Iberoamerican Mathematical Olympiad (by its initials in Spanish)
APMO	Asian Pacific Mathematical Olympiad
OMCC	Centroamerica and the Caribbean Mathematical Olympiad (by its initials in Spanish)
OMM	Mexican Mathematical Olympiad (by its initials in Spanish)
USAMO	USA Mathematical Olympiad
(Country, year)	The problem was used in the olympiad of that country in one of its stages and the corresponding year

Further Reading

1. Pérez Seguí, M. L., *Combinatoria, Cuadernos de Olimpiadas de Matemáticas*. Instituto de Matemáticas, UNAM, 2000.
2. Riordan, J., *Introduction to Combinatorial Analysis*. Dover, Mineola, 2002.
3. Andreescu, T. and Feng, Z., *A Path to Combinatorics for Undergraduates*. Birkhäuser, Basel, 2004.
4. Andreescu, T. and Feng, Z., *102 Combinatorial Problems*. Birkhäuser, Boston, 2003.
5. Anderson, I., *A First Course in Combinatorial Mathematics*, 2nd edition. Oxford University Press, London, 1989.
6. Stanley, R. P., *Enumerative Combinatorics*, 2nd edition, Vol. 1. Cambridge University Press, Cambridge, 2011.

P. Soberón, *Problem-Solving Methods in Combinatorics*,
DOI 10.1007/978-3-0348-0597-1, © Springer Basel 2013

Index

P. Soberón, *Problem-Solving Methods in Combinatorics*,
DOI 10.1007/978-3-0348-0597-1, © Springer Basel 2013

Made in the USA
Coppell, TX
08 January 2025